Bridge Safety, Maintenance and Management in a Life-Cycle Context

Dan M. Frangopol
Department of Civil and Environmental Engineering
ATLSS Engineering Research Center
Lehigh University, Bethlehem, Pennsylvania
USA

Sunyong Kim
Department of Civil and Environmental Engineering
Wonkwang University, Jeonbuk
Republic of Korea

CRC Press
Taylor & Francis Group
Boca Raton London New York

CRC Press is an imprint of the
Taylor & Francis Group, an **informa** business

A SCIENCE PUBLISHERS BOOK

Cover credit: The cover photo is the Akashi-Kaikyo Bridge. This is a stiffened truss suspension bridge with the longest span in the world, located in Akashi Strait connecting by highway the city of Kobe and the Awaji Island in Japan. The central and total lengths of its three spans are 1,991m and 3,991m, respectively. The authors would like to thank Dr. Kiyohiro Imai, Senior Director of the Corporate Planning Department, Honshu-Shikoku Bridge Expressway Company, Ltd., Kobe, Japan, for providing the photo.

First edition published 2022
by CRC Press
6000 Broken Sound Parkway NW, Suite 300, Boca Raton, FL 33487-2742

and by CRC Press
2 Park Square, Milton Park, Abingdon, Oxon, OX14 4RN

© 2022 Taylor & Francis Group, LLC

CRC Press is an imprint of Taylor & Francis Group, LLC

Library of Congress Cataloging-in-Publication Data
Names: Frangopol, Dan M., author. | Kim, Sunyong, 1976- author.
Title: Bridge safety, maintenance and management in a life-cycle context / Dan M. Frangopol,
    Department of Civil and Environmental Engineering, Atlss Engineering Research Center, Lehigh
    University, Bethlehem, United States, Sunyong Kim, Department of Civil and Environmental
    Engineering, Wonkwang University, Jeonbuk Republic of Korea.
Description: First edition. | Boca Raton, FL : CRC Press, 2022. | Includes bibliographical references
    and index. | Summary: "This book presents state-of-the-art theoretical background and practical
    applications for bridge safety, maintenance and management in a life-cycle context. The primary
    topics of this book include bridge safety and service life prediction, bridge inspection and
    maintenance, bridge resilience and sustainability, and life-cycle performance and cost analysis for
    individual bridges and bridge networks under hazards and climate change. This book can help
    students, researchers and practitioners to build up knowledge on life-cycle bridge performance
    and cost management at both individual bridge level and network level under various deteriorating
    effects, hazards and climate change"-- Provided by publisher.
Identifiers: LCCN 2021035267 | ISBN 9781032052816 (hbk) | ISBN 9781032052847 (pbk) |
    ISBN 9781003196877 (ebk)
Subjects: LCSH: Bridges--Maintenance and repair. | Product life cycle.
Classification: LCC TG315 .F74 2022 | DDC 624.2--dc23
LC record available at https://lccn.loc.gov/2021035267

ISBN: 978-1-032-05281-6 (hbk)
ISBN: 978-1-032-05284-7 (pbk)
ISBN: 978-1-003-19687-7 (ebk)

DOI: 10.1201/9781003196877

Typeset in Times New Roman
by Shubham Creation

Preface

Bridges, one of the most critical components of a nation's infrastructure, have a positive socio-economic impact. Among more than 617,000 bridges in the United States, at least 42% exceed 50 years of their service life, and 7.5% of the existing bridges are classified as structurally deficient. Even though the number of structurally deficient bridges in the United States has declined recently, the number of bridges in fair condition has continuously increased. Along with increasing the number of structurally deficient bridges in many countries, demands for efficiently and effectively managing deteriorating bridges and extending their service life have increased. In general, substantial financial resources are needed to maintain or increase the safety and performance of deteriorating bridges and to extend their service life. Under limited financial resources, bridge owners and managers have to consider the life-cycle cost and performance to make optimum decisions to both maximize bridge performance and minimize the costs of inspection, repair, rehabilitation and replacement in a life-cycle context. These cost-effective actions should consider the uncertainties involved in the life-cycle performance and cost of bridges and bridge networks including their progressive and/or sudden deteriorations under extreme events, environmental factors, and climate change.

During the past two decades, it has been generally acknowledged that life-cycle bridge analysis can be a systematic tool to address efficient and effective bridge management under uncertainty, life-cycle management at the bridge network level can lead to an improvement in the allocation of limited financial resources ensuring the safety and functionality of the bridge network, life-cycle management of bridges and bridge networks based on resilience and sustainability can improve their resistance and robustness to extreme events such as earthquakes, tsunamis, floods, and hurricanes, and bridge management should consider the impact of environmental conditions and climate change.

In this book, important concepts and approaches developed recently on bridge safety, maintenance, and management in a life-cycle context are systematically addressed. Bridge life-cycle performance and cost analysis, prediction, optimization, and decision making under uncertainty are discussed. The major topics include bridge safety and service life prediction, bridge inspection and structural health monitoring, bridge maintenance, life-cycle bridge and bridge network management, optimum life-cycle bridge management planning, resilience and sustainability of bridges and bridge networks under hazards, and bridge management considering climate change. By providing practical applications of the presented concepts and approaches, this book can help students, researchers, practitioners, infrastructure owners and managers, and transportation officials to build up their knowledge of life-cycle bridge performance and cost management at both project level and network level under various deteriorating mechanisms, hazards, and climate change effects.

This book includes ten chapters. Chapter 1 provides the basic probability concepts and methods associated with bridge life-cycle performance and cost analysis under uncertainty. The general concepts of bridge life-cycle performance and cost analysis based on the deterministic and probabilistic approaches are described. The reliability of structural components and systems, lifetime distributions, extreme value distribution, exceedance probability, structural performance indicators, and importance of the components in a structural system are provided.

Chapter 2 deals with the time-dependent structural performance and service life of deteriorating bridges under uncertainty. The effects of the structural performance prediction with maintenance on bridge life-cycle performance and service life are described. The updating process with new information from inspection and structural health monitoring (SHM) and its integration into bridge life-cycle performance and service life prediction are described. The effects of inspection and maintenance on the performance and service life of deteriorating bridges are provided.

Chapter 3 presents several representative inspection and SHM techniques for bridge management. The probabilistic approaches to predict the bridge reliability using the data from monitoring and to quantify the availability of monitoring data are provided. The formulations of the damage detection time, damage detection delay, and damage detection-based probability of failure are presented. These formulations can be used for optimum life-cycle inspection, monitoring, and maintenance management planning.

Chapter 4 addresses the effects of bridge maintenance on performance, service life, service life extension, and cost. These effects are modeled using the reliability profiles, lifetime functions, damage propagation, and event trees. The effect of correlation among the service life extensions due to maintenance on the extended service life of components and systems is investigated. The bridge maintenance process and bridge management systems used in several countries are reported.

Chapter 5 describes the general procedure for life-cycle performance and cost analysis under uncertainty, and, in this context, presents the probabilistic approach for optimum bridge design and management planning. The multi-objective optimization and multi-attribute decision making are provided, where the correlation among objectives, and essential and redundant objectives are investigated.

Chapter 6 provides applications for optimum life-cycle bridge inspection and management planning. The applications are categorized into three topics: probabilistic bridge performance and service life prediction, probabilistic optimum bridge inspection and monitoring planning, and probabilistic optimum bridge maintenance planning.

Chapter 7 deals with the life-cycle bridge network management approach under uncertainty. The probabilistic performance indicators used for bridge network management are introduced. These indicators consider connectivity, travel flow, cost, and risk of the bridge network. The multi-objective optimum life-cycle bridge network management and the associated applications are described. The bridge network management presented in this chapter considers the progressive performance deterioration of bridges under normal loading effects and environmental conditions.

Chapter 8 provides the general concept of the resilience of bridges and bridge networks. The effects of bridge performance deterioration and hazard-induced damage, and risk mitigation on resilience are addressed. The representative functionality recovery models are provided. Furthermore, the approaches to assess sustainability are provided. Representative applications of resilience and sustainability for bridge and bridge network management are summarized.

Chapter 9 deals with the impact of climate change on bridge performance and representative adaptation measures to reduce this impact. The concepts of benefit-cost analysis and gain-loss analysis, and their applications are presented. Using these analyses, the adaptation types and times for bridge and bridge network management under climate change are optimized.

Chapter 10 provides the summary of this book, representative conclusions of each chapter, and future efforts in the research fields of life-cycle bridge safety, maintenance, and management in a life-cycle context.

The cover photo is the Akashi-Kaikyo Bridge. This is a stiffened truss suspension bridge with the longest span in the world, located in Akashi Strait connecting by highway the city of Kobe and the Awaji Island in Japan. The central and total lengths of its three spans are 1,991m and 3,991m, respectively. The bridge construction took 10 years to complete from 1988 to 1998. The central span was originally designed to be 1,990m, but the Kobe earthquake of 1995 forced the two towers, which were still under construction, one meter farther apart. The two towers of the bridge reach 298m above sea level. The cover image was provided by Dr. Kiyohiro Imai, Senior Director of the Corporate Planning Department, Honshu-Shikoku Bridge Expressway Company, Ltd., Kobe, Japan.

Dan M. Frangopol
Department of Civil and Environmental Engineering
ATLSS Engineering Research Center, Lehigh University
117 ATLSS Drive, Bethlehem, PA 18015-4729, USA
Email: dan.frangopol@lehigh.edu

Sunyong Kim
Department of Civil and Environmental Engineering
Wonkwang University, 460 Iksandae-ro Iksan,
Jeonbuk, 54538, Republic of Korea
Email: sunyongkim@wku.ac.kr

Contents

Chapter 1

Probabilistic Concepts and Methods for Bridge Life-Cycle Analysis

NOTATIONS (continued)	
p_{exd}	= exceedance probability
p_f	= probability of failure
$p_{f,dmg}$	= probability of failure of a damaged system
$p_{f,int}$	= probability of failure of an intact system
P_{in}	= initial structural performance
P_{life}	= lifetime structural performance
p_s	= reliability
P_{th}	= threshold of structural performance
$Q(t)$	= time-variant functionality of a bridge under a hazard
Q_{dmg}	= ultimate load capacity of a damaged structure
Q_{fun}	= applied load when the structure cannot serve its intended purpose due to large deformation and significant damage
Q_{ini}	= initial functionality
Q_{int}	= ultimate load capacity of an intact structural system
Q_n	= nominal load effect
Q_U	= ultimate load
r_{dmg}	= particular damage state
r_{int}	= pristine system state
r_p	= deterioration rate
$r_{p,v}$	= deterioration rate after preventive maintenance
$R_{K,dir}$	= direct risk
$R_{K,ind}$	= indirect risk
R_n	= nominal strength
R_{dmf}	= redundancy factor for the collapse limit state
R_f	= redundancy factor
R_{fun}	= redundancy factor for the functionality limit state
R_{rsd}	= residual strength factor
R_{rsv}	= reserve strength factor
R_{ult}	= redundancy factor for the ultimate collapse state
R_U	= plastic resistance of a structure
R_K	= risk
RB	= robustness index
RB_{risk}	= risk-based robustness index
RI	= redundancy index
RS	= functionality-based resilience
$S(t)$	= survival function
SF	= safety factor
t_{ini}	= damage initiation time
$t_{ini,e}$	= damage initiation time after essential maintenance
$t_{life,0}$	= initial service life
$t_{life,em}$	= extended service life with essential maintenance
$t_{life,em+pm}$	= extended service life with both essential maintenance and preventive maintenance
t_{ob}	= observed time period
t_{pm}	= application time for preventive maintenance

NOTATIONS (continued)	
$t_{pm,e}$	= damage propagation delay after preventive maintenance
$Y_{max,n}$	= maximum value among n variables
$Y_{min,n}$	= minimum value among n variables
VI	= vulnerability index
β	= reliability index
β_{dmg}	= reliability index of a damaged system
β_{int}	= reliability index of an intact system
ΔP_{em}	= structural performance improvement after essential maintenance
ΔP_{pm}	= structural performance improvement after preventive maintenance
γ_i	= load factor for the ith load case
μ_X	= mean of X
ϕ	= resistance factor
ϕ_j	= structural function of the jth system state
Φ	= standard normal cumulative distribution
ν	= mean occurrence rate of the events per unit time
σ_{all}	= allowable stress
σ_{max}	= maximum stress induced by loading
σ_X	= standard deviation of X
Γ	= performance index considering the total travel time and distance

ABSTRACT

A review of the basic probability concepts and methods used in reliability analysis is provided in Chapter 1. The bridge life-cycle performance and cost based on deterministic and probabilistic concepts and methods are described. The reliability of structural components and systems, lifetime distributions, extreme value distribution, and exceedance probability are introduced. Furthermore, the structural performance indicators for bridge life-cycle performance and cost analysis are described. Finally, probabilistic approaches to determine the importance of the components in structural systems are presented for efficient bridge management at both individual bridge system-level and bridge network-level. The concepts and methods presented in this chapter are helpful to understand the following chapters.

1.1 INTRODUCTION

Bridge life-cycle performance and cost analysis and prediction require consideration of time-dependent effects including loading, resistance deterioration, maintenance and repair, among others (Frangopol 2011, 2018; Frangopol and Soliman 2016). The deterioration process due to corrosion and fatigue causes a progressive reduction in the performance of existing bridges. Extreme events such as earthquakes, hurricanes, floods, and fires produce a sudden drop in bridge performance (Biondini and Frangopol 2016). The change in the structural performance after maintenance actions needs to be investigated taking into account maintenance types, times of maintenance applications, and available financial resources (Sánchez-Silva et al. 2016). The performance prediction and cost estimation of new and existing bridges are highly uncertain (Frangopol et al. 2012). For this reason, bridge life-cycle performance and cost analysis require probabilistic concepts and methods.

This chapter presents a review of the basic probability theory used in reliability analysis in order to understand bridge life-cycle performance and cost analysis under uncertainty. The general relation between bridge life-cycle performance and cost is described using both deterministic and probabilistic concepts and methods. The failure or survival of structural components and systems, lifetime distributions, extreme value distribution, and exceedance probability are introduced. The structural performance indicators used for bridge life-cycle analysis are described. Finally, probabilistic approaches to determine the importance of the components in structural systems are presented for efficient bridge management at both individual bridge system-level and bridge network-level.

1.2   BRIDGE LIFE-CYCLE PERFORMANCE AND COST ANALYSIS UNDER UNCERTAINTY

The structural performance of deteriorating bridges during their life-cycle is continuously affected by external loadings, mechanical stressors, environmental conditions, and extreme events, among others. Therefore, the deterioration of the structural performance may be unavoidable if no maintenance actions are applied. The maintenance actions on bridges are applied for: (a) maintaining the structural safety level above a prescribed threshold level; (b) improving the structural performance; and (c) extending the service life of deteriorating bridges (Frangopol et al. 2001). However, application of the maintenance actions requires expenses. Under limited financial resources, the cost for maintenance has to be allocated in a rational way. For this reason, it is necessary to determine the types of maintenance actions and the times of their applications by using systematic approaches. These approaches should consider the effects of maintenance actions on structural performance and cost under uncertainty (Kong and Frangopol 2003a, 2003b).

1.2.1   Structural Performance Profiles under Uncertainty

Figure 1.1 shows the general representation of the lifetime bridge performance profile with essential and preventive maintenance actions. The essential maintenance (EM), performance-based intervention, is applied when the structural performance reaches the predefined threshold P_{th}. The preventive maintenance (PM) is applied at a scheduled time, and leads to a delay in the deterioration process. As shown in Figure 1.1, the improvement in structural performance after the EM action is larger than the structural performance improvement after the application of PM (Frangopol and Kim 2011, 2014b). The uncertainties associated with the initial performance, deterioration initiation, deterioration rate, duration of service life without maintenance, effect of preventive maintenance, effect of essential maintenance, service life with preventive maintenance, and service life with both preventive and essential maintenance are indicated by using probability density distributions (PDFs). Figure 1.2 shows a schematic multi-linear representation of the nonlinear life-cycle performance profile shown in Figure 1.1, considering application of EM before the application of PM. The uncertainties associated with initial structural performance P_{in}, damage initiation time t_{ini}, damage initiation time after EM $t_{ini,e}$, initial service life $t_{life,0}$, deteriorating rate r_p, deterioration rate after PM $r_{p,v}$, structural performance improvement after EM ΔP_{em}, structural performance improvement after PM ΔP_{pm}, damage propagation delay after PM $t_{pm,e}$, extended service life with EM $t_{life,em}$, and extended service life with both EM and PM $t_{life,em+pm}$ are represented by probability density functions (PDFs) as shown in Figure 1.2(b). The parameters and PDFs related to the time-dependent structural performance profiles can be determined based on predicting the damage occurrence and propagation, investigating the relation between the damage propagation and the structural performance, and estimating the effect of maintenance actions on the structural performance.

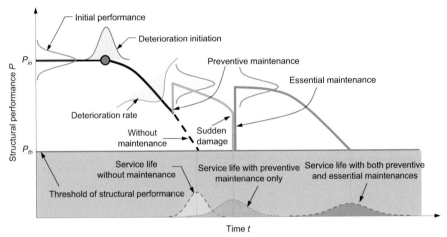

Figure 1.1   General lifetime bridge performance profile (adapted from Frangopol 2011).

Figure 1.2 is based on the assumption that the preventive maintenance action is applied after damage initiation. Depending on the relation between the damage initiation time t_{ini} and the time of first preventive maintenance action $t_{pm,1}$, three cases are considered as follows:

- Case 1: structural performance deterioration initiates before the first preventive maintenance application (i.e., $t_{ini} \leq t_{pm,1}$)
- Case 2: structural performance deterioration initiates within the duration of the first preventive maintenance effect (i.e., $t_{pm,1} \leq t_{ini} \leq t_{pm,1} + t_{pm,e}$)
- Case 3: structural performance deterioration initiates after the duration of the first preventive maintenance effect (i.e., $t_{pm,1} + t_{pm,e} \leq t_{ini}$)

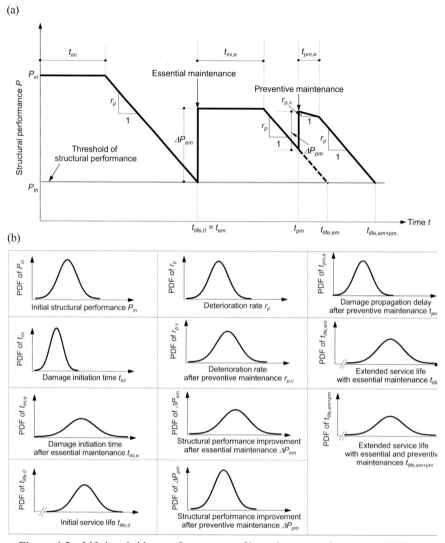

Figure 1.2   Lifetime bridge performance profile under uncertainty: (a) multi-linear profile; (b) PDFs of the random variables representing the multi-linear profile.

The structural performance profiles associated with cases 1, 2 and 3 are shown in Figure 1.3. The associated formulations of these three cases are provided in Frangopol et al. (2001). In this manner, the structural performance for multiple preventive and essential maintenance can be computed by using a simulation method (e.g., Monte Carlo simulation, Latin Hypercube sampling). Based on the structural performance profiles with the preventive and essential maintenance under uncertainty, the expected structural performance level at a predefined time, expected extended service life, expected remaining service life, and associated PDFs can be obtained.

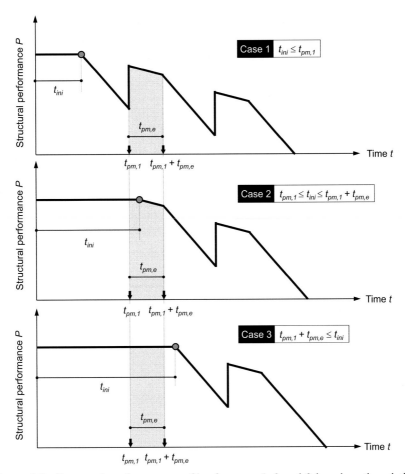

Figure 1.3 Structural performance profiles for cases 1, 2 and 3 based on the relation between the first preventive maintenance application time and deterioration initiation time.

1.2.2 Life-Cycle Cost Profiles under Uncertainty

The life-cycle cost of a bridge consists of the initial cost, costs associated with inspection and maintenance actions, and failure cost. Accordingly, the total life-cycle cost C_{life} during a predefined lifetime t^*_{life} can be expressed as (Frangopol et al. 1997b)

$$C_{life} = C_{ini} + C_{ins} + C_{ma} + C_{fail} \qquad (1.1)$$

where C_{ini} = initial cost including design and construction cost; C_{ins} = cost of inspection; C_{ma} = cost of maintenance; and C_{fail} = expected cost of failure. C_{fail} is estimated as

$$C_{fail} = C_f \times p_{f,life} \qquad (1.2)$$

where C_f is the expected monetary loss induced by structural failure, and $p_{f,life}$ is the lifetime probability of failure. The costs C_{ins}, C_{ma} and C_{fail} in Eq. (1.1) are affected by the threshold of structural performance P_{th} shown in Figures 1.1 and 1.2.

In order to keep the performance of an existing bridge above a larger threshold level P_{th}, the inspection and maintenance cost $C_{ins} + C_{ma}$ has to increase. By increasing the threshold of the bridge performance P_{th}, the lifetime probability of failure $p_{f,life}$ can be reduced, and therefore, expected failure cost C_{fail} can decrease. Furthermore, a reduction in the expected failure cost C_{fail} can be obtained by increasing the inspection and maintenance cost $C_{ins} + C_{ma}$. The effects of the threshold of structural performance on the total life-cycle cost C_{life} are shown in Figure 1.4, where the optimum threshold of structural performance P^*_{th} is associated with the minimum present value of expected cost C^*_{life}.

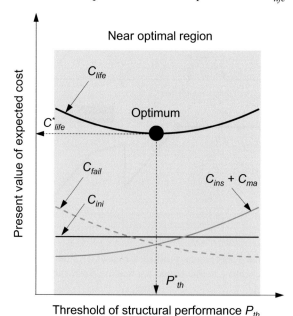

Figure 1.4   Relation between total life-cycle cost and threshold of structural performance of an existing bridge in optimal region.

When a new bridge is designed and constructed, its lifetime structural performance P_{life} can affect the costs C_{ini}, C_{ins}, C_{ma} and C_{fail}. The mean value of the structural performance or minimum structural performance during a predefined lifetime t^*_{life} can be treated as the lifetime structural performance. When the bridge is designed and constructed to have a larger lifetime structural performance, a larger initial cost can be required. However, the inspection and maintenance cost $C_{ins} + C_{ma}$ and expected failure cost C_{fail} may be reduced. Therefore, it can be expected that a larger initial cost C_{ini} produces both less inspection and maintenance cost $C_{ins} + C_{ma}$ and less failure cost C_{fail}. As a result, the relation between total life-cycle cost C_{life} and lifetime structural performance P_{life} in optimal region can be obtained as shown in Figure 1.5. The minimum present value of expected cost C^*_{life} corresponds to the optimum lifetime structural performance P^*_{life}. It should be noted that the initial cost C_{ini} in Figure 1.4 is a fixed value, since this cost is associated with a given existing bridge. However, the initial cost C_{ini} in Figure 1.5 is treated as a variable. This is because new bridges can be designed and constructed considering the relation between the lifetime structural performance P_{life} and initial cost C_{ini}.

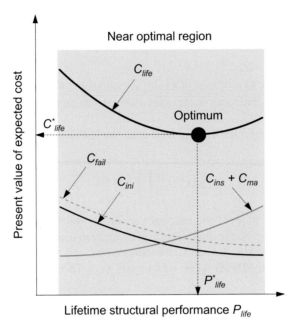

Figure 1.5 Relation between total life-cycle cost and lifetime structural performance of a new bridge in optimal region.

1.3 PROBABILITY CONCEPTS AND METHODS FOR BRIDGE LIFE-CYCLE ANALYSIS

In order to assess and predict the bridge life-cycle performance and cost under uncertainty, probabilistic concepts and methods need to be used. In this section, the reliability of structural components and systems, lifetime distributions, modeling of extreme values, and estimation of exceedance probability are introduced.

1.3.1 Structural Reliability

The reliability of a structure indicates the probability that the structure will fulfill its intended purpose or survive under a specific time interval and environmental conditions (Ang and Tang 1984; Melchers 1999; Leemis 2009). The reliability of a bridge p_s is generally computed based on the state function $g(\mathbf{X})$ expressed as

$$g(\mathbf{X}) = R(\mathbf{X}) - S(\mathbf{X}) \tag{1.3}$$

where $\mathbf{X}$ is the vector of random variables (i.e., $\mathbf{X} = \{X_1, X_2, \ldots, X_n\}$), and R and S represent the functions associated with the resistance and load effect, respectively. Accordingly, the reliability p_s and probability of failure p_f are defined as

$$p_s = P\big(g(\mathbf{X}) \geq 0\big) = \int_{g(\mathbf{X})>0} f_{\mathbf{X}}(\mathbf{x})\, d\mathbf{x} \tag{1.4a}$$

$$p_f = 1 - p_s = P\big(g(\mathbf{X}) < 0\big) = \int_{g(\mathbf{X})<0} f_{\mathbf{X}}(\mathbf{x})\, d\mathbf{x} \tag{1.4b}$$

where $[g(\mathbf{X}) > 0]$, $[g(\mathbf{X}) < 0]$ and $[g(\mathbf{X}) = 0]$ are associated with safe, failure and limit state, respectively, and $f_{\mathbf{X}}(\mathbf{x})$ is the joint PDF of the random variables $\mathbf{X} = \{X_1, X_2, \ldots, X_n\}$.

If the random variables $\mathbf{X}$ are independent, the joint PDF $f_{\mathbf{X}}(\mathbf{x})$ can be expressed as

$$f_{\mathbf{X}}(\mathbf{x}) = \prod_{i=1}^{n} f_{X_i}(x_i) \tag{1.5}$$

where $f_{X_i}(x_i)$ is the PDF of the random variable X_i. In this case, the solution of Eq. (1.4) may be obtained using the general integration method. For example, when the two random variables X_1 and X_2 are independent, and the state function is $g(\mathbf{X}) = X_1 - X_2$, the joint PDF is $f_{\mathbf{X}}(\mathbf{x}) = f_{X_1}(x_1) \cdot f_{X_2}(x_2)$. Therefore, the probability of failure p_f becomes

$$p_f = \int_{-\infty}^{\infty} \int_{-\infty}^{x_1 < x_2} f_{X_1}(x_1) f_{X_2}(x_2)\, dx_1 dx_2 \tag{1.6}$$

where X_1 and X_2 are associated with the resistance and load effect, respectively. Figure 1.6 shows the PDFs of random variables X_1 and X_2, and the probability of failure. As shown in Figure 1.6(a), when X_2 is deterministic (i.e., the outcome takes only a single value $X_2 = X_2{}^*$), the area under the PDF $f_{X_1}(x_1)$ upper bounded by $X_2{}^*$ is the probability of failure. Also, if X_1 is deterministic (i.e., the outcome takes only a single value $X_1 = X_1{}^*$), the probability of failure is represented by the area under the PDF $f_{X_2}(x_2)$ lower bounded by $X_1{}^*$ as shown in Figure 1.6(b). Considering that the cumulative distribution function (CDF) of a random variable X is

$$F_X(x) = \int_{-\infty}^{x} f_X(y)\, dy \quad \text{for } x \geq y \tag{1.7}$$

Eq. (1.6) can be also expressed as (Melchers 1999)

$$p_f = \int_{-\infty}^{\infty} F_{X_1}(x) f_{X_2}(x)\, dx \tag{1.8a}$$

$$p_f = \int_{-\infty}^{\infty} \big[1 - F_{X_2}(x)\big] f_{X_1}(x)\, dx \tag{1.8b}$$

which is the sum of the probability of failure for all X_2 when X_2 exceeds X_1 (see Eq. (1.8a)), or for all X_1 when X_2 exceeds X_1 (see Eq. (1.8b)). This case is shown in Figure 1.6(c).

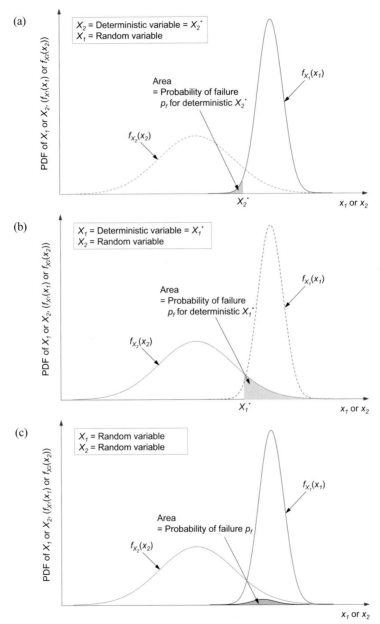

Figure 1.6 PDFs of X_1 and X_2, and probability of failure: (a) probability of failure for deterministic X_2^*; (b) probability of failure for deterministic X_1^*; (c) probability of failure associated with random variables X_1 and X_2.

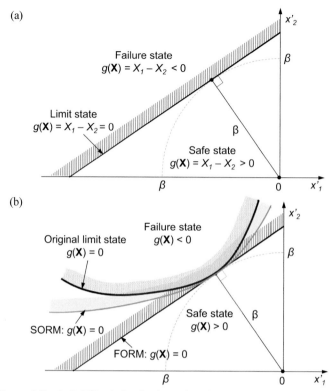

Figure 1.7   Reliability index in the reduced space: (a) linear state function; (b) nonlinear state function.

When the two random variables X_1 and X_2 are normally distributed with means μ_{X_1} and μ_{X_2}, and standard deviations σ_{X_1} and σ_{X_2}, respectively, the probability of failure p_f can be expressed with the state function $g(\mathbf{X}) = X_1 - X_2$ as (Ang and Tang 1984)

$$p_f = \Phi\left(-\frac{\mu_M}{\sigma_M}\right) = \Phi\left[\frac{-\left(\mu_{X_1} - \mu_{X_2}\right)}{\sqrt{\sigma_{X_1}^2 + \sigma_{X_2}^2}}\right] = \Phi(-\beta) \qquad (1.9)$$

where Φ is the standard normal cumulative distribution, and μ_M and σ_M are the mean and standard deviation of $X_1 - X_2$, respectively. The reliability index β can be expressed as

$$\beta = \Phi^{-1}(p_s) = \Phi^{-1}(1 - p_f) \qquad (1.10)$$

where Φ^{-1} is the inverse of the standard normal cumulative distribution. Consider the reduced variables of X_1 and X_2 defined, respectively, as

$$X_1' = \frac{X_1 - \mu_{X_1}}{\sigma_{X_1}} \qquad (1.11a)$$

$$X_2' = \frac{X_2 - \mu_{X_2}}{\sigma_{X_2}} \qquad (1.11b)$$

In the space of the two reduced variables X_1' and X_2', the minimum distance from the origin to $g(\mathbf{X}) = X_1 - X_2 = 0$ is the reliability index β as shown in Figure 1.7(a). When the state function $g(\mathbf{X})$ is nonlinear, the limit state can be expressed as the first order or second order equation to find the minimum distance from the origin. Figure 1.7(b) compares the safe, failure and limit states when the limit state $g(\mathbf{X}) = 0$ represented by the first order (FORM) or second order reliability method (SORM) are applied. It should be noted that Eqs. (1.5) to (1.11) can be applied for independent random variables. When the correlations among the random variables $\mathbf{X}$ are considered, the multi-dimensional integration for Eq. (1.4) can be performed by using numerical approximation methods such as Monte Carlo simulation and first/second-order reliability methods (FORM/SORM).

1.3.2 Reliability of Structural Systems

A structural system is composed of multiple components. The reliability of a structural system can be computed using the reliability of its components. In general, a structural system can be modeled based on series, parallel and series-parallel systems. Figure 1.8 shows examples of series, parallel and series-parallel systems consisting of three components. The probability of failure p_f of each of the five systems shown can be expressed as

$$p_f = P(F_1 \cup F_2 \cup F_3) \quad \text{for a series system} \qquad (1.12a)$$

$$p_f = P(F_1 \cap F_2 \cap F_3) \quad \text{for a parallel system} \qquad (1.12b)$$

$$p_f = P[F_1 \cup (F_2 \cap F_3)] \quad \text{for a series-parallel system with component 1} \quad (1.12c)$$
in series

$$p_f = P[F_2 \cup (F_1 \cap F_3)] \quad \text{for a series-parallel system with component 2} \quad (1.12d)$$
in series

$$p_f = P[F_3 \cup (F_1 \cap F_2)] \quad \text{for a series-parallel system with component 3} \quad (1.12e)$$
in series

where F_i is the failure event of the *i*th component, and $\cup$ and $\cap$ represent union and intersection, respectively. Based on the state function $g(\mathbf{X})$, the probability of failure p_f for series, parallel and series-parallel systems can be formulated as

$$p_f = P\left(\bigcup_{i=1}^{3} \{ g_i(\mathbf{X}) \le 0 \} \right) \qquad \text{for a series system} \quad (1.13a)$$

$$p_f = P\left(\bigcap_{i=1}^{3} \{ g_i(\mathbf{X}) \le 0 \} \right) \qquad \begin{array}{l} \text{for a parallel} \\ \text{system} \end{array} \quad (1.13b)$$

$$p_f = P\left(\left[g_1(\mathbf{X}) \le 0\right] \cup \left[\{g_2(\mathbf{X}) \le 0\} \cap \{g_3(\mathbf{X}) \le 0\}\right]\right)$$ for a series-parallel system with component 1 in series (1.13c)

$$p_f = P\left(\left[g_2(\mathbf{X}) \le 0\right] \cup \left[\{g_1(\mathbf{X}) \le 0\} \cap \{g_3(\mathbf{X}) \le 0\}\right]\right)$$ for a series-parallel system with component 2 in series (1.13d)

$$p_f = P\left(\left[g_3(\mathbf{X}) \le 0\right] \cup \left[\{g_1(\mathbf{X}) \le 0\} \cap \{g_2(\mathbf{X}) \le 0\}\right]\right)$$ for a series-parallel system with component 3 in series (1.13e)

where $g_i(\mathbf{X}) \le 0$ represents the failure state of the ith component. If the components of the system are perfectly correlated, the probabilities of failure p_f of the series and parallel systems can be computed as

$$p_f = \max_{i=1}^{n}\{p_{f,i}\}$$ for a series system   (1.14a)

$$p_f = \min_{i=1}^{n}\{p_{f,i}\}$$ for a parallel system   (1.14b)

where $p_{f,i}$ is the probability of failure of the ith component. For series and parallel systems consisting of independent components, p_f is

$$p_f = 1 - \prod_{i=1}^{n}\left(1 - p_{f,i}\right)$$ for a series system   (1.15a)

$$p_f = \prod_{i=1}^{n} p_{f,i}$$ for a parallel system   (1.15b)

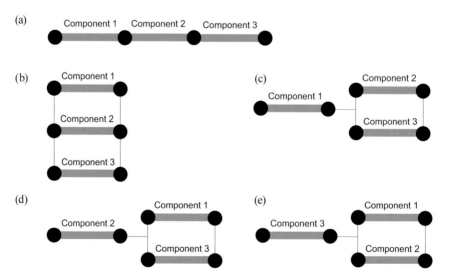

Figure 1.8 Structural system consisting of three components: (a) series system; (b) parallel system; (c) series-parallel system with component 1 in series; (d) series-parallel system with component 2 in series; (e) series-parallel system with component 3 in series.

For more complex systems considering the correlation among the random variables and types of distributions associated with each of the random variables involved in the computation process, the failure probability p_f and reliability $p_s = 1 - p_f$ can be computed using several software programs such as CALREL (Liu et al. 1989), STRUREL (Rackwitz 1996), PROBAN (Bjerager 1996), RELSYS (Estes and Frangopol 1998), OpenSees (Der Kiureghian et al. 2006), COSSAN-X (Patelli 2017), FERUM v4.0 (Bourinet et al. 2009), UQlab (Marelli and Sudret 2014), and UQpy (Shields et al. 2019). Furthermore, the probability of failure p_f or reliability p_s at time t can be computed by using the time-dependent state function $g(\mathbf{X}(t))$, where $\mathbf{X}(t)$ is the vector of time-dependent random variables (Mori and Ellingwood 1994a, 1994b; Enright and Frangopol 1998b, 1998c, 2000; Gong and Frangopol 2019).

1.3.3 Lifetime Distributions

Lifetime distributions including survival function, hazard function, cumulative hazard function, and mean residual life function are useful for bridge life-cycle performance prediction, since the closed-form of the reliability of a bridge system can be formulated, and time-dependent reliability can be computed efficiently (Leemis 2009; Okasha and Frangopol 2010a and 2010b; Barone and Frangopol 2013). The survival function $S(t)$ is defined as the probability that a structure survives at any time t as (Leemis 2009)

$$S(t) = P\left[T_f \geq t\right] = \int_t^\infty f(x)\,dx \qquad \text{for } t \geq 0 \qquad (1.16)$$

where T_f is the time to failure of the structure, and the associated PDF is denoted by $f(x)$. The survival function $S(t)$ has to satisfy three conditions: (a) $S(0) = 1$; (b) $\lim_{t \to \infty} S(t) = 0$; (c) $S(t)$ monotonically decreases with time t. When a large number of bridges in a network have the same lifetime distribution, the survival function $S(t)$ indicates the expected fraction of the bridges surviving at time t. The survival function $S(t)$ is the complementary cumulative distribution function (CDF) (i.e., $S(t) = 1 - F(t)$, where $F(t) = P[T_f < t]$). Therefore, the PDF $f(t)$ can be expressed as

$$f(t) = -S'(t) = \lim_{\Delta t \to 0} \frac{P[t \leq T_f \leq t + \Delta t]}{\Delta t} \qquad (1.17)$$

where $S'(t)$ is the derivative of the survival function $S(t)$. The availability $A(t)$ is the probability that a component is functioning at time t (Ang and Tang 1984). It should be noted that the survival function $S(t)$ is the probability that a component will not fail before time t as indicated in Eq. (1.16). Therefore, the survival function $S(t)$ for a non-repairable component is the same as the availability $A(t)$. However, when the component is repaired or replaced, the formulations of $S(t)$ and $A(t)$ may be different (Okasha and Frangopol 2010b). Additional details will be described subsequently.

The hazard function $h(t)$, representing the instantaneous rate of failure at time t, is defined as the ratio of the PDF $f(t)$ to the survival function $S(t)$ as (Leemis 2009).

$$h(t) = -\frac{S'(t)}{S(t)} = \frac{f(t)}{S(t)} \tag{1.18}$$

There are two conditions for $h(t)$ to be satisfied: (a) $\int_0^\infty h(t)\,dt = \infty$; (b) $h(t) \geq 0$ for $t \geq 0$.

The cumulative hazard function $H(t)$ represents the accumulated risk until the time t, which is expressed as (Leemis 2009)

$$H(t) = \int_o^t h(x)\,dx \tag{1.19}$$

$H(t)$ has to satisfy three conditions: (a) $H(0) = 0$; (b) $\lim_{t \to \infty} H(t) = \infty$; (c) $H(t)$ monotonically increases.

Furthermore, the mean residual life function $L(t)$ is the expected remaining life (i.e., $T_f - t$) given that a structure has survived until the time t (i.e., $T_f \geq t$), which is given by (Leemis 1986)

$$L(t) = E\left[T_f - t \middle| T_f \geq t\right] = \frac{1}{S(t)} \int_t^\infty xf(x)dx - t \tag{1.20}$$

The formulation of $L(t)$ has to be based on the three conditions: (a) $L(t) \geq 0$; (b) $L'(t) \geq -1$; (c) $\int_0^\infty \frac{1}{L(t)}\,dt = \infty$.

The most appropriate distribution type to represent the time to failure of deteriorating bridge components or system is the Weibull distribution (Okasha and Frangopol 2010a, 2010b). The PDF of the Weibull distribution $f(t)$ is

$$f(t) = \kappa\lambda(\lambda t)^{\kappa-1} \cdot \exp\left[-(\lambda t)^\kappa\right] \qquad \text{for } t \geq 0 \tag{1.21}$$

where κ and λ are the shape and scale parameters, respectively. According to Eq. (1.17), the survival function $S(t)$ based on the Weibull distribution is

$$S(t) = \exp\left[-(\lambda t)^\kappa\right] \tag{1.22}$$

In this case, the hazard function $h(t)$ defined in Eq. (1.18) can be formulated as

$$h(t) = \frac{f(t)}{S(t)} = \kappa\lambda(\lambda t)^{\kappa-1} \tag{1.23}$$

Therefore, based on Eq. (1.19), the cumulative hazard function $H(t)$ becomes

$$H(t) = (\lambda t)^\kappa \tag{1.24}$$

Finally, the mean residual life function $L(t)$ based on the Weibull distribution can be obtained using Eq. (1.20). The survival function $S(t)$ of Eq. (1.22), hazard function $h(t)$ of Eq. (1.23), cumulative hazard function $H(t)$ of Eq. (1.24) and the mean residual life function $L(t)$ are illustrated in Figure 1.9, where the Weibull distribution (see Eq. 1.21) is used for the components 1, 2 and 3. The shape factors κ for the components 1, 2 and 3 are assumed as 0.5, 1.0 and 2.0, respectively. The same scale parameter $\lambda = 0.1$ is assumed for these three components.

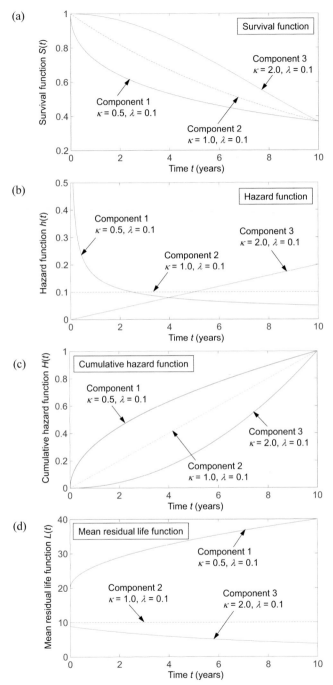

Figure 1.9   Lifetime distributions based on Weibull PDF: (a) survival function $S(t)$; (b) hazard function $h(t)$; (c) cumulative hazard function $H(t)$; (d) mean residual life function $L(t)$.

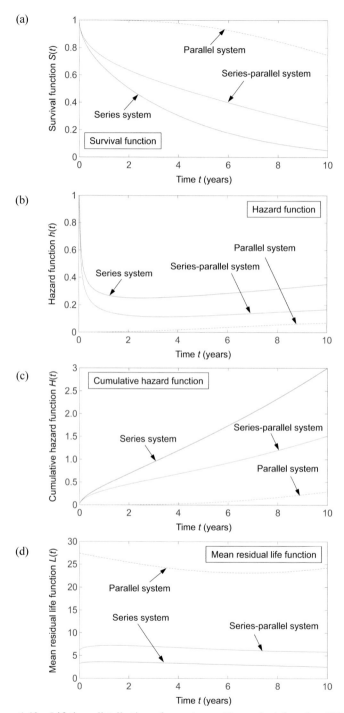

Figure 1.10 Lifetime distributions for systems: (a) survival function $S(t)$; (b) hazard function $h(t)$; (c) cumulative hazard function $H(t)$; (d) mean residual life function $L(t)$.

The lifetime distribution for a system can be formulated based on the component survival functions. For example, when the three components shown in Figure 1.8 are independent, the survival function for the series (Figure 1.8(a)), parallel (Figure 1.8(b)) and series-parallel (Figure 1.8(c)) systems can be expressed using Eq. (1.15) as

$$S_{series}(t) = S_1(t) \cdot S_2(t) \cdot S_3(t) \qquad \text{for a series system} \qquad (1.25a)$$

$$S_{paral}(t) = 1 - [1 - S_1(t)] \cdot [1 - S_2(t)] \cdot [1 - S_3(t)] \quad \text{for a parallel system} \qquad (1.25b)$$

$$S_{comb}(t) = S_1(t) \cdot \left[1 - \{1 - S_2(t)\} \cdot \{1 - S_3(t)\} \right] \quad \begin{array}{l}\text{for a series-parallel}\\ \text{system with}\\ \text{component 1 in series}\end{array} \qquad (1.25c)$$

where $S_1(t)$, $S_2(t)$ and $S_3(t)$ are the survival functions of the components 1, 2 and 3, respectively. These functions are shown in Figure 1.9(a). The survival functions $S(t)$, hazard functions $h(t)$, cumulative hazard functions $H(t)$, and mean residual life functions $L(t)$ of the three systems are shown in Figure 1.10. The series system has the smallest $S(t)$ and $L(t)$, and the largest $h(t)$ and $H(t)$. Conversely, the largest $S(t)$ and $L(t)$, and the smallest $h(t)$ and $H(t)$ are obtained for the parallel system.

1.3.4 Extreme Value Distributions

Extreme values (i.e., largest and smallest values) associated with load effect and resistance are significant for assessment, prediction and management of the bridge performance and service life as well as bridge design. The extreme values need to be treated as random variables, since they are outcomes of the structural analyses and observed measurements under uncertainty (Gumbel 1958, Ang and Tang 1984, 2007). The formulation of the distribution of the extreme value is based on the distribution of the initial variables X_1, X_2, ..., X_n. Assuming that the initial variables are identically distributed [i.e., $F_{X_1}(x) = F_{X_2}(x) = ... = F_{Xn}(x) = F_X(x)$] and statistically independent, the CDF of $Y_{max,n}$ is expressed as

$$Y_{max,n}(y) = P(Y_{max,n} \le y) = P(X_1 \le y, X_2 \le y,...,X_n \le y) = [F_X(y)]^n \quad (1.26)$$

where $Y_{max,n} = max\{X_1, X_2, ..., X_n\}$. Similarly, the CDF of the smallest value $Y_{min,n}$ is

$$Y_{min,n}(y) = 1 - P(Y_{min,n} > y) = 1 - P(X_1 > y, X_2 > y,...,X_n > y)$$
$$= 1 - [1 - F_X(y)]^n \qquad (1.27)$$

where $Y_{min,n} = min\{X_1, X_2, ..., X_n\}$. As the sample size n increases substantially, tending to infinity, the asymptotic CDFs of the largest and smallest values may converge to a particular distribution type. The types of the asymptotic CDFs are affected by each end of tail's behavior of the initial distribution. According to Gumbel (1958), the asymptotic CDF can be categorized into three types: (a) the double exponential form (i.e., Type I asymptotic form), (b) the exponential form (i.e., Type II asymptotic form), and (c) the exponential form with upper or lower bound (i.e., Type III asymptotic form).

For example, the largest and smallest values of the initial variables with normal and exponential distributions having exponential tails correspond to the double exponential form. The asymptotic CDF of the double exponential form can be expressed as

$$F_{Y_{max}}(y) = \exp\left[-\exp[-\gamma_{max}(y-\alpha_{max})]\right] \qquad \text{for the largest value} \qquad (1.28a)$$

$$F_{Y_{min}}(y) = 1-\exp\left[-\exp[-\gamma_{min}(y-\alpha_{min})]\right] \quad \text{for the smallest value} \qquad (1.28a)$$

where α_{max} and α_{min} are the location parameters, and γ_{max} and γ_{min} are the scale parameters. Figure 1.11 compares the exact and asymptotic CDFs, which are based on Eqs. (1.26) and (1.28a), respectively, when the number of sample size n from a standard normal distribution is equal to 10, 100 and 1000. It is shown that the difference between the exact and asymptotic CDFs is reduced by increasing the number of samples n.

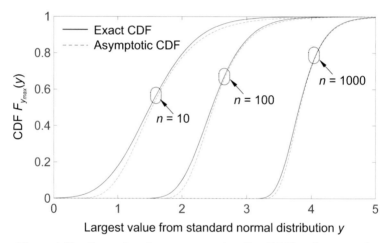

Figure 1.11   Comparison between exact (see Eq. (1.26)) and asymptotic (see Eq. (1.28a)) CDFs for the largest value from standard normal distribution.

Furthermore, the largest value of lognormal distribution corresponds to the Type II asymptotic form. The Type III asymptotic distribution can represent the extreme values of uniform and triangular distributions, which are bounded by an upper or a lower limit (Ang and Tang 1984, 2007). The Weibull distribution defined in Eq. (1.21) belongs to the Type III asymptotic form. The time to failure of a bridge is generally represented by the Weibull distribution. The Type I, II and III asymptotic forms can be generalized as (Resnick 1987; Coles 2001)

$$F(y) = \exp\left[-\left\{1+\xi\left(\frac{y-\alpha}{\gamma}\right)\right\}^{-1/\xi}\right] \qquad (1.29)$$

where ξ is the shape parameter, α is the location parameter, and γ is the scale parameter.

1.3.5   Exceedance Probability

Exceedance probability has been applied to predict the occurrence of extreme events such as earthquakes, hurricanes and floods (Leadbetter et al. 1983; Embrechts et al. 1997). The Poisson process can provide the probability associated with the number of the future occurrences in a given time interval between 0 and t, which is defined as (Melchers 1999; Ang and Tang 2007)

$$P(N_e = n_e) = \frac{(vt)^{n_e}}{n_e!} \cdot \exp[-vt] \qquad (1.30)$$

where n_e = number of the event occurrences; and v = mean occurrence rate of the events per unit time. The probability that the number of occurrences will be at least one is computed as

$$P(N_e \geq 1) = 1 - P(N_e = 0) = 1 - \exp[-vt] \qquad (1.31)$$

If the initial variable X follows the Type I or II asymptotic distribution, the probability that the largest value among N future observations $Y_{max,N}$ exceeds the largest value among n existing observations $Y_{max,n}$ is estimated as (Ang and Tang 1984)

$$P(Y_{max,N} > Y_{max,n}) = 1 - \exp\left[-\frac{N}{n}\right] \qquad (1.32)$$

where $Y_{max,n} = \{X_1, X_2, ..., X_n\}$. When the initial variable X is the largest annual value, the probability that the value in the future years t exceeds the largest value during the observed years t_{ob} can be computed by modifying Eq. (1.32) as

$$P_{exd} = 1 - \exp\left[-\frac{t}{t_{ob}}\right] \qquad (1.33)$$

which is the same as Eq. (1.31), since the mean occurrence rate v in Eq. (1.31) is equal to $1/t_{ob}$. Furthermore, the probability that the value in the future years t exceeds the largest value during the observed years t_{ob}, or the value in the future years t is less than the smallest value during the observed years t_{ob}, is computed as

$$P_{exd} = 1 - \exp\left[-\frac{2t}{t_{ob}}\right] \qquad (1.34)$$

1.4   STRUCTURAL PERFORMANCE INDICATORS

Life-cycle bridge management requires to assess and predict the structural performance under uncertainty (Frangopol and Soliman 2016; Frangopol et al. 2017). The performances of individual components and structural systems are represented by adequate indicators reflecting structural condition, safety, tolerance to damage under normal conditions and extreme events (Ghosn et al. 2016a, 2016b). The condition-based performance indicators depend on the outcomes of the visual inspection, where the degree of damage (e.g., section loss due to corrosion and size of fatigue crack) is inspected. In the United States, the bridge condition measurement is rated with National Bridge Inventory (NBI) condition rating and AASHTOWare Bridge Management™ Condition State (CS) (or PONTIS). For the NBI condition rating method, the conditions of bridge components such as deck, super- and sub-structures are rated using a value from 0 to 9. The condition rating values of 0 and 9 indicate the failure condition and excellent condition, respectively. The CS of AASHTOWare Bridge Management™ has the rating values from 1 to 5, where no evidence of damage in a bridge component is associated with the rating value of 1 and the severe damage of a bridge component corresponds to the value of 5. More information on these two condition rating methods and their applications is available in FHWA (1995), CDOT (1998), Estes and Frangopol (2003), Cambridge Systematics, Inc (2009), AASHTO (2011, 2015), Saydam et al. (2013a), and Frangopol and Saydam (2014).

The safety-based structural performance indicators include the safety margin M_S, probability of failure p_f, probability of survival (or reliability) p_s, reliability index β, survival function $S(t)$, hazard function $h(t)$, and cumulative hazard function $H(t)$. The formulations of p_f and p_s are indicated in Eq. (1.4). Eqs. (1.9) and (1.10) show the relation among p_f, p_s, and β. The definitions of $S(t)$, $h(t)$ and $H(t)$ are provided in Eqs. (1.16), (1.18) and (1.19), respectively. The structural performance indicators p_f, p_s, β, $S(t)$, $h(t)$ and $H(t)$ can be applied to both structural components and systems.

Furthermore, the safety factor (SF) can be used as a performance indicator (FHWA 2013a). Based on the allowable stress design (ASD), the SF is expressed as

$$SF_{ASD} = \frac{\sigma_{max}}{\sigma_{all}} \qquad (1.35)$$

where σ_{max} = maximum stress induced by loading (e.g., yield stress, buckling stress); and σ_{all} = allowable stress. According to the load and resistance factor design (LRFD) method, SF can be computed as (Ghosn et al. 2016b)

$$SF_{LRFD} = \frac{\phi R_n}{\sum \gamma_i Q_{n,i}} \qquad (1.36)$$

where ϕ is the resistance factor, R_n is the nominal strength, and γ_i is the load factor associated with the nominal load effect $Q_{n,i}$ for the ith load case. In the plastic analysis considering the collapse load multiplier, the SF can be estimated as (FHWA 2013a)

$$SF_{PLST} = \frac{R_U}{Q_U} = \frac{R_U}{\lambda_D Q_D + \lambda_L Q_L} \qquad (1.37)$$

where R_U is the plastic resistance of the structure, and Q_U is the ultimate load. λ_D and λ_L are the collapse load multipliers for the dead load and live load, respectively.

The structural redundancy, vulnerability and robustness are involved in the reliability-based system performance indicator (Ghosn et al. 2016a). The structural redundancy evaluates the reserve and residual strength of the damaged bridge system. The reserve strength factor R_{rsv} is the ratio of the ultimate load capacity of an intact structural system Q_{int} to the applied load Q as (Frangopol and Curley 1987)

$$R_{rsv} = \frac{Q_{int}}{Q} \tag{1.38}$$

When the applied load Q is zero, the reserve strength factor R_{rsv} has a value of infinity. The residual strength factor R_{rsd} is defined as (Frangopol and Curley 1987; Frangopol and Klisinski 1989)

$$R_{rsd} = \frac{Q_{dmg}}{Q_{int}} \tag{1.39}$$

where Q_{dmg} = ultimate load capacity of the damaged structure. The factor $R_{rsd} = 0$, when the damaged structure collapses (i.e., $Q_{dmg} = 0$). If there is no change in load-carrying capacity before and after damage (i.e., $Q_{dmg} = Q_{int}$), $R_{rsd} = 1$. The redundancy factor R_f is expressed as (Frangopol and Curley 1987; Frangopol and Klisinski 1989)

$$R_f = \frac{1}{1 - R_{rsd}} \tag{1.40}$$

Furthermore, considering loss of the cross-sectional area of a bridge component, the damage factor D_f is given as (Cohn 1980)

$$D_f = 1 - \frac{A_d}{A} \tag{1.41}$$

where A_d = damaged cross-sectional area; and A = intact cross-sectional area.

The deterministic redundancy factors for different limit states can be defined as (Ghosn and Moses 1998; Ghosn et al. 2016b)

$$R_{ult} = \frac{Q_{int}}{Q_{mem}} \tag{1.42a}$$

$$R_{fun} = \frac{Q_{fun}}{Q_{mem}} \tag{1.42b}$$

$$R_{dmf} = \frac{Q_{dmg}}{Q_{mem}} \tag{1.42c}$$

where R_{ult}, R_{fun} and R_{dmf} are the redundancy factors for the ultimate collapse state, functionality limit state, and collapse limit state of the damaged structure, respectively. Q_{mem} is the applied load associated with the first member failure, Q_{int} is the ultimate load capacity of the intact structure, and Q_{fun} is the applied load when the structure cannot serve its intended purpose due to large deformation and significant damage. Figure 1.12 shows the values of Q_{mem}, Q_{int}, Q_{fun} and Q_{dmg}.

The structural redundancy index RI is expressed as (Frangopol and Curley 1987; Frangopol and Nakib 1991; FHWA 2013a)

$$RI_d = \beta_{int} - \beta_{dmg} \tag{1.43a}$$

$$RI_\beta = \frac{\beta_{int}}{\beta_{int} - \beta_{dmg}} \tag{1.43b}$$

$$RI_f = \frac{p_{f,dmg} - p_{f,int}}{p_{f,int}} \tag{1.43c}$$

where RI_d and RI_β are the reliability index-based redundancy index, and RI_f is the probability failure-based redundancy index. β_{int} is the reliability index of the intact bridge system and β_{dmg} is the reliability index of the damaged bridge system. $p_{f,dmg}$ is the probability of failure of the damaged bridge system and $p_{f,int}$ is the probability of failure of the intact bridge system. According to Eq. (1.43a), a larger difference between β_{int} and β_{dmg} causes a larger redundancy index RI_d. For the redundancy index RI_β indicated in Eq. (1.43b), a larger β_{int} and/or smaller difference between the reliability indices before and after damage occurrence result in a larger redundancy index RI_β. As indicated in Eq. (1.43c), a larger redundancy index RI_f is caused by a smaller $p_{f,int}$ and/or larger difference between $p_{f,dmg}$ and $p_{f,int}$. The difference between $p_{f,dmg}$ and $p_{f,int}$ is associated with the availability of warning before system failure. Furthermore, based on the availability of a structural system, the redundancy index at time t can be computed as (Okasha and Frangopol 2010a)

$$RI_a = \frac{A_s(t) - A_w(t)}{1 - A_s(t)} = \frac{UA_w(t) - UA_s(t)}{UA_s(t)} \tag{1.44}$$

where A_s = availability of the system; and A_w = availability of the weakest components. UA_s and UA_w are the unavailabilities of the system and the weakest component, respectively. The sum of availability and unavailability is equal to one (i.e., $A_s + UA_s = 1$ or $A_w + UA_w = 1$).

The vulnerability index (VI) is defined as the ratio of the probability of failure of the damaged bridge system to the failure probability of the intact bridge system as (Lind 1995)

$$VI = \frac{p_f(r_{dmg}, S)}{p_f(r_{int}, S)} \tag{1.45}$$

where p_f is the probability of failure of the bridge system, r_{dmg} is a particular damage state, r_{int} is a pristine system state, and S is the prospective loading.

The damage tolerance index DI represents the reciprocal of VI (FHWA 2013a) as

$$DI = \frac{1}{VI} = \frac{p_f(r_{int}, S)}{p_f(r_{dmg}, S)} \tag{1.46}$$

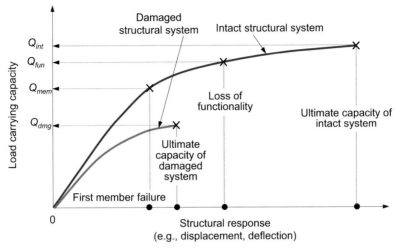

Figure 1.12   Load carrying capacities Q_{mem}, Q_{int}, Q_{fun} and Q_{dmg} of intact and damaged structural systems (adapted from Ghosn et al. 2010).

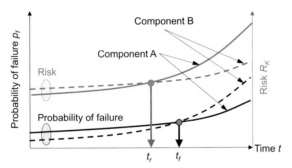

Figure 1.13   Probability of failure and risk profiles of two deteriorating structural components.

The structural robustness indicates the tolerance to damage resulted from the extreme or accidental loads (Ghosn et al. 2010; Saydam and Frangopol 2011; Frangopol and Saydam 2014). Maes et al. (2006) defined the structural robustness *RB* of a bridge system as

$$RB = \min_{i} \frac{p_{f,int}}{p_{f,i}} \tag{1.47}$$

where $p_{f,i}$ is the probability of failure of a bridge system when its *i*th component is damaged.

The reliability-based structural performance indicators such as probability of survival, reliability index, redundancy, vulnerability and robustness reflect the uncertainties associated with resistance, load effects and structural modeling. These indicators are associated with the safety of a structure with respect to specific limit states, and do not address the consequences (i.e., monetary losses) resulting from a failure event. The risk as a structural performance indicator R_K is formulated by combining the occurrence probability of a specific event with the resulting monetary loss (Zhu and Frangopol 2013a; Dong and Frangopol 2016a). The probability of failure and risk profiles of two deteriorating structural components A and B are compared in Figure 1.13. Both the probability of failure and risk increase over time. The probability of failure p_f of component A is higher than p_f of component B before time t_f. However, the risk R_K of component A is smaller than R_K of component B before time t_r, where t_r is less than t_f ($t_r < t_f$), as shown in Figure 1.13. This is because a component with higher probability of failure may have a smaller monetary loss resulting from its failure, compared to another component with lower failure probability but higher monetary loss. Figure 1.14 shows illustrative risk profiles of a deteriorating bridge with essential maintenance, where the seismic and traffic risks are considered. The essential maintenance is applied twice during the lifetime of the bridge when the total risk is close to the risk threshold.

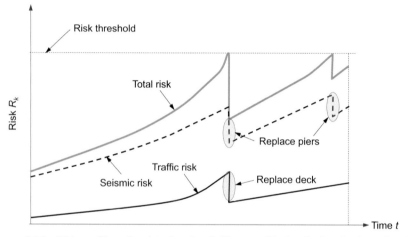

Figure 1.14   Risk profiles of a deteriorating bridge considering both seismic and traffic risks (adapted from Zhu and Frangopol 2013a).

The monetary losses are categorized into direct and indirect losses (Saydam et al. 2013b). The risk R_K is the sum of the direct risk $R_{K,dir}$ and indirect risk $R_{K,ind}$. The formulations of $R_{K,dir}$ and $R_{K,ind}$ are given as (Zhu and Frangopol 2012)

$$R_{K,dir} = p_{f,i} \, C_{dir,i} \tag{1.48a}$$

$$R_{K,ind} = p_{f,i} \, p_{f,sub|i} \, C_{ind,i} \tag{1.48b}$$

in which $p_{f,i}$ is the probability of the failure of the ith component, $p_{f,sub|i}$ is the probability of system failure due to the failure of the ith component, and $C_{dir,i}$ and $C_{ind,i}$ are the direct and indirect monetary losses caused by the failure of component i, respectively. The direct consequence $C_{dir,i}$ can be estimated as the cost to replace the failed component i, and the indirect consequence $C_{ind,i}$ consists of the costs to rebuild the system, safety loss and environmental loss (Saydam et al. 2013a). Furthermore, the risk-based robustness index RB_{risk} can be expressed as (Baker et al. 2008)

$$RB_{risk} = \frac{R_{K,dir}}{R_{K,dir} + R_{K,ind}} \tag{1.49}$$

The risk-based robustness index RB_{risk} of Eq. (1.49) ranges from zero to one. If the indirect risk $R_{K,ind}$ does not significantly affect the total risk $R_{K,dir} + R_{K,ind}$, the value of RB_{risk} will be close to one, and the associated system is called a robust system.

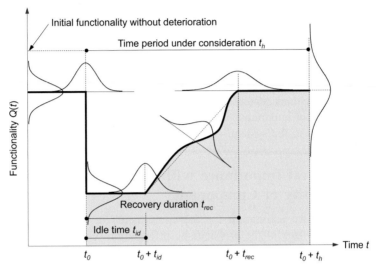

Figure 1.15   Functionality recovery profile under uncertainty (adapted from Decò et al. 2013).

Resilient bridge structures can reduce the consequences caused by failure and the recovery time (FHWA 2013a). Accordingly, the resilience can be quantified by considering (a) functionality during and after extreme events; (b) time to recover the structural performance level before extreme events; (c) cost and impact on community when loss of functionality occurs (Tierney and Bruneau 2007; Bocchini and Frangopol 2012a; Lounis and McAllister 2016). The functionality-based resilience RS at the investigated time t_h can be defined as (Frangopol and Bocchini 2011; Decò et al. 2013)

$$RS = \frac{\int_{t_0}^{t_0+t_h} Q(t)dt}{t_h \cdot Q_{ini}} \tag{1.50}$$

where t_0 is the time of hazard occurrence, $Q(t)$ is the time-variant functionality of a bridge under a hazard, and Q_{ini} is the initial functionality. The functionality recovery profile associated with Eq. (1.50) is presented in Figure 1.15. The concept of resilience can be applied for various infrastructure systems including power supply networks, transportation networks, and communication systems (Lounis and McAllister 2016). For the bridge network system, the functionality $Q(t)$ is formulated as (Frangopol and Bocchini 2011; Bocchini and Frangopol 2012a)

$$Q(t) = \frac{\Gamma(t) - \Gamma^0}{\Gamma^{100} - \Gamma^0} \tag{1.51}$$

where $\Gamma(t)$ is the performance index considering the total travel time and distance, and Γ^0 and Γ^{100} are the performance indices when all the bridges in the network are in service and out of service, respectively.

1.5 STRUCTURAL COMPONENT IMPORTANCE

The components of the bridge have different contributions to the structural system performance (Rausand and Hoyland 2004; Modarres et al. 2017). By estimating the contributions of the components to the entire bridge system, efficient and effective bridge maintenance management can be achieved (Liu and Frangopol 2005c). The structural component importance can be estimated with and without the reliability of the components in the system.

1.5.1 Structural Importance without Considering Reliability of Component

When the reliability estimation for the components in the bridge system is unavailable due to lack of the probabilistic information related to resistance and load effect, the structural importance factor of the ith component of the bridge system $I_{s,i}$ can be computed as (Leemis 2009)

$$I_{s,i} = \sum_{j=1}^{2^{n-1}} \left[\phi_j \left(\mathbf{x} \big| x_i = 1 \right) - \phi_j \left(\mathbf{x} \big| x_i = 0 \right) \right] \tag{1.52}$$

where n is the number of components in the bridge system. The safe and failure states of the ith component correspond to $x_i = 1$ and $x_i = 0$, respectively. The bridge system state vector $\mathbf{x}$ consists of the states of the components (i.e., $\mathbf{x} = \{x_1, x_2,..., x_n\}$). The structural function of the jth bridge system state ϕ_j is defined as

$$\phi_j(\mathbf{x}) = \begin{cases} 1 & \text{when the bridge system is in a safe state} \\ 0 & \text{when the bridge system is in a failure state} \end{cases} \tag{1.53}$$

$\phi_j(\mathbf{x}|x_i = 1)$ in Eq. (1.52) indicates the structural function of the jth system state when the ith component is in a safe state (i.e., $x_i = 1$). Given that the ith component is in a failure state (i.e., $x_i = 0$), the structural function of the jth system is denoted as $\phi_j(\mathbf{x}|x_i = 0)$ in Eq. (1.52).

1.5.2 Component Importance based on Reliability

Considering that the change in the reliability of the most important component results in the largest change in the system reliability, the reliability-based importance of the ith component is defined as (Birnbaum 1969)

$$I_{b,i} = \frac{\partial p_{s,sys}}{\partial p_{s,i}} \quad (1.54)$$

where $I_{b,i}$ is the Birnbaum importance factor of component i, $p_{s,sys}$ is the system reliability, and $p_{s,i}$ is the reliability of the ith component. When the components in the system are independent, the Birnbaum importance factor $I_{b,i}$ can be computed as (Rausand and Hoyland 2004)

$$I_{b,i} = p_{s,sys}(\mathbf{p_s}|p_{s,i} = 1) - p_{s,sys}(\mathbf{p_s}|p_{s,i} = 0) \quad (1.55)$$

where $\mathbf{p_s}$ is the vector consisting of the reliabilities of n components in the system (i.e., $\mathbf{p_s} = \{p_{s,1}, p_{s,2}, \ldots, p_{s,n}\}$), and $p_{s,sys}(\mathbf{p_s}|p_{s,i} = 1)$ and $p_{s,sys}(\mathbf{p_s}|p_{s,i} = 0)$ are the system reliability for given reliability of the ith component $p_{s,i}$ equal to 1 and 0, respectively.

It should be noted that the Birnbaum importance factor $I_{b,i}$ of Eq. (1.55) is computed based on the system reliability and reliability of the other components except the ith component. In order to reflect the reliability of the component i for estimating the importance factor, the criticality importance factor $I_{cr,i}$ of the component i can be used as (Modarres et al. 2017)

$$I_{cr,i} = I_{b,i} \frac{p_{s,i}}{p_{s,sys}} \quad (1.56)$$

Furthermore, risk achievement worth (RAW) importance factor $I_{raw,i}$ and risk reduction worth (RRW) importance factor $I_{rrw,i}$ of the component i can be defined, respectively, as (Ericson 2015)

$$I_{raw,i} = p_{s,sys} - p_{s,sys}(\mathbf{p_s}|p_{s,i} = 0) \quad (1.57a)$$

$$I_{rrw,i} = p_{s,sys}(\mathbf{p_s}|p_{s,i} = 1) - p_{s,sys} \quad (1.57b)$$

Since the scales of the importance factors provided in Eqs. (1.52) and (1.53)–(1.57) are not consistent, the importance factors need to be normalized as

$$I_i^{norm} = \frac{I_i}{\sum\limits_{j=1}^{n} I_j} \quad (1.58)$$

Figures 1.16, 1.17 and 1.18 show the normalized importance factors of the three components of the series, parallel and series-parallel system, respectively, shown in Figures 1.8(a), 1.8(b) and 1.8(c), respectively. Herein, the survival functions $S(t)$ of the three components defined in Figure 1.9(a) are treated as the reliabilities p_s. For the series system in Figure 1.8(a), the three components have the same normalized structural, critical and RAW importance factors [i.e., $I_s^{norm} = I_{cr}^{norm} = I_{raw}^{norm} = 0.333$] as shown in Figure 1.16(a). The component 1 associated with the smallest $S(t)$ [see Figure 1.9(a)] has the largest Birnbaum and RRW importance factors among the three components [see Figures 1.16(b) and 1.16(c)]. However, the component 1 of the parallel system has the smallest Birnbaum,

critical and RAW importance factors among the three components as shown in
Figure 1.17(b) and 1.17(c). Figures 1.18(a) and 1.18(d) show that the component 1
has the largest normalized importance factor when the structural, RRW importance
factors are applied. It should be noted that the normalized structural importance
factors $I_{s,i}^{norm}$ are not affected by the reliability of component as indicated in
Eq. (1.52), and therefore, are constant over time as shown in Figures 1.16(a),
1.17(a) and 1.18(a).

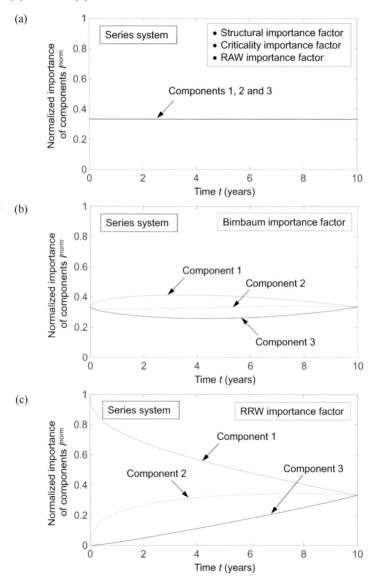

Figure 1.16 Normalized importance factors of components for the series system shown
in Figure 1.8(a): (a) structural importance factor $I_{s,i}^{norm}$; (b) Birnbaum importance factor
$I_{b,i}^{norm}$; (c) risk reduction worth importance factor $I_{rrw,i}^{norm}$.

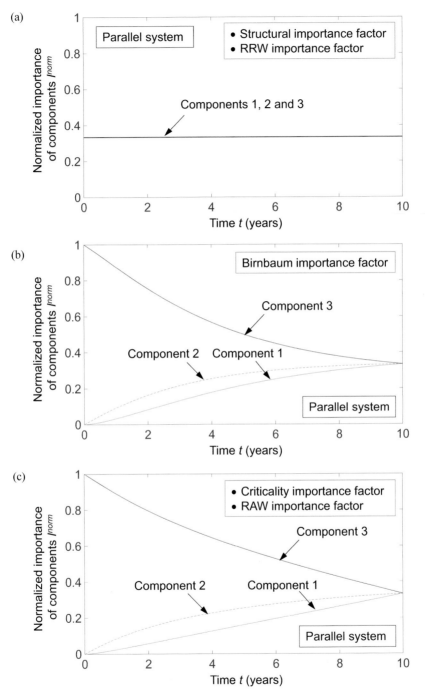

Figure 1.17 Normalized importance factors of components for the parallel system shown in Figure 1.8(b): (a) structural importance factor $I_{s,i}^{norm}$; (b) Birnbaum importance factor $I_{b,i}^{norm}$; (c) criticality importance factor $I_{cr,i}^{norm}$.

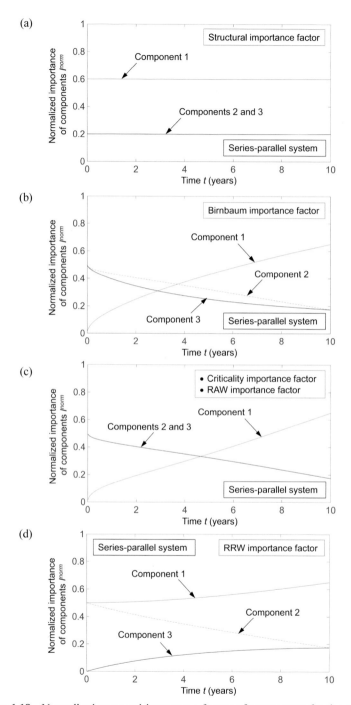

Figure 1.18 Normalized structural importance factors of components for the series-parallel system shown in Figure 1.8(c): (a) structural importance factor $I_{s,i}^{norm}$; (b) Birnbaum importance factor $I_{b,i}^{norm}$; (c) criticality importance factor $I_{cr,i}^{norm}$; (d) RRW importance factor $I_{rrw,i}^{norm}$.

1.6   CONCLUDING REMARKS

The main purpose of this chapter is to provide the background information associated with bridge life-cycle performance and cost analysis under uncertainty. The general concepts of bridge life-cycle performance and cost based on the deterministic and probabilistic approaches are described. The fundamental probabilistic concepts and methods used in the bridge life-cycle are presented. The failure probability or reliability of structural components and systems, lifetime distributions, probabilistic expressions for extreme values, and exceedance probability are addressed. Furthermore, the structural performance indicators for bridge life-cycle performance and cost analysis are described. Finally, the importance of the components in a structural system is presented. The concepts and methods described in this chapter are helpful to understand the following chapters.

Chapter **2**

Structural Performance and Service Life

a	= crack size
a_{crt}	= critical crack size
a_0	= initial crack size
A_{re}	= remaining area of the reinforcement
C_0	= chloride concentration at surface
C_{loss}	= unit monetary loss
C_{spl}	= unit cost for sampling
d_s	= diameter of reinforcement
$d(t)$	= damage intensity at time t
D	= effective chloride diffusion coefficient
D_f	= Miner's damage accumulation index
E_{loss}	= expected monetary loss
f_{loss}	= loss function
$f_X(x)$	= probability density function of random variable X
$f'_X(x)$	= initial (i.e. prior) probability density function of random variable X
$f''_X(x)$	= updated probability density function of random variable X
$g(t)$	= state function at time t
$L(x)$	= likelihood function
M_L	= moment by applied loads
M_R	= flexural capacity
M_s	= safety margin
n_{spl}	= number of samples
N_{an}	= annual number of cycles
p_f	= probability of failure
p_s	= reliability
P_{det}	= probability of damage detection
P_{in}	= initial structural performance

NOTATIONS (continued)	
P_{th}	= threshold of structural performance
$P(t)$	= maximum pit depth at time t
r_{an}	= increase rate of annual number of cycles
r_{cor}	= corrosion rate
R	= ratio of maximum pit depth to uniform pit depth
S_{re}	= equivalent stress range
t_{crt}	= time to reach critical crack size
t_{insp}	= inspection time
t_{life}	= service life
$t_{life,0}$	= initial service life
$t_{life,ex}$	= expected extended service life
t_{cor}	= corrosion initiation time
$t_{life,up}$	= updated service life
V_L	= shear force by applied loads
V_R	= shear strength
$Y(a)$	= geometry function for crack size a
β	= reliability index
ΔK	= stress intensity factor

ABSTRACT

Chapter 2 deals with the time-dependent structural performance and service life of deteriorating bridges under uncertainty. The uncertainties associated with bridge performance and service life prediction are described. The updating process with new information from inspection and structural health monitoring and its integration into bridge life-cycle performance and service life prediction is presented. The time-dependent bridge deterioration processes such as corrosion and fatigue are described. The probabilistic bridge performance is predicted by considering the state functions of the structural components and systems and lifetime distributions. Finally, the effects of inspection and maintenance on bridge performance and service life under uncertainty are investigated.

2.1 INTRODUCTION

The life-cycle performance and cost of deteriorating bridges have been investigated extensively (Frangopol 2011; Frangopol et al. 2017). However, there are difficulties in these investigations, which are mainly caused by the uncertainties involved in life-cycle structural performance, life-cycle cost and service life prediction. Appropriate understanding and quantification of these uncertainties will lead to more efficient and effective service life management. Furthermore, it is necessary to investigate the effects of inspection and maintenance on the structural performance and service life for optimum life-cycle management of deteriorating bridges (Frangopol and Soliman 2016; Frangopol et al. 2017; Frangopol 2018; Frangopol and Kim 2019).

This chapter investigates the time-dependent structural performance and service life of deteriorating bridges under uncertainty. The uncertainties associated with bridge performance and service life prediction are described. In order to estimate the monetary loss resulted from untimely and/or inappropriate maintenance actions of deteriorating bridges, the loss function is presented. The updating process to integrate new information from inspection and structural health monitoring (SHM) into bridge life-cycle performance and service life prediction is described. The time-dependent bridge deterioration processes such as corrosion and fatigue are reviewed. Furthermore, the formulations of the state functions based on the corrosion and fatigue damage are presented. These formulations can be used to compute the probabilistic structural performance indicators. Finally, the effects of inspection and maintenance on bridge performance and service life are investigated.

2.2   BRIDGE SERVICE LIFE UNDER UNCERTAINTY

The prediction of structural performance and remaining service life serves as the basis for life-cycle performance and cost analysis, and optimum service life management of deteriorating bridges. In general, the progressive and sudden drop of the structural performance and remaining service life is time-dependent under uncertainty (Dong et al. 2013; Akiyama et al. 2014; Biondini and Frangopol 2016; Frangopol and Soliman 2016). For this reason, a proper understanding and estimation of uncertainties associated with bridge performance and service life are essential (Frangopol and Kim 2014a, 2019).

2.2.1   Uncertainties in Bridge Performance and Service Life Prediction

The bridge performance can be represented by the structural performance indicators (e.g., reliability, risk, redundancy, robustness, vulnerability) described in Section 1.4. The service life can be defined as the time to reach a predefined structural performance threshold (Frangopol and Kim 2014a, 2019). The various parameters and models involved in structural performance and service life prediction are affected by uncertainties. These uncertainties can be categorized into aleatory and epistemic (Ang and Tang 2007; Herrman 2015). The aleatory uncertainty is associated with natural randomness. This uncertainty is not reducible. The epistemic uncertainty is associated with inaccuracies in the prediction and estimation of reality, and can be reduced by improving the knowledge and data related to bridge performance and service life prediction.

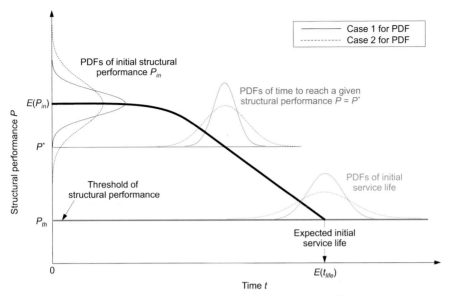

Figure 2.1 Time-dependent structural performance and service life under uncertainty.

Figure 2.1 shows the PDFs of the initial structural performance, time for reaching a prescribed structural performance P^*, and initial service life considering two cases for the variance (i.e., square of the standard deviation). In case 2, the variance is larger than in case 1. It can be seen that the variance of the initial structural performance P_{in} increases together with the variance of the initial service life t_{life}. Also, the variance associated with performance increases over time. The effect of uncertainties associated with the safety margin M_s on the probability of failure p_f is shown in Figure 2.2, where the PDFs of the safety margins $M_{s,1}$ and $M_{s,2}$ have

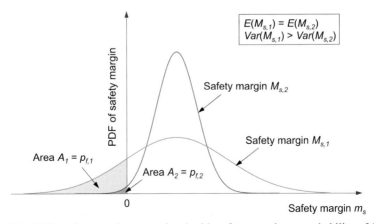

Figure 2.2 Effect of uncertainty associated with safety margin on probability of failure.

the same expected values [i.e., $E(M_{s,1}) = E(M_{s,2})$], but the variance of $M_{s,1}$ is larger than that of $M_{s,2}$ [i.e., $Var(M_{s,1}) > Var(M_{s,2})$]. As a result, the probability of failure

$p_{f,1}$ for $M_{s,1}$ is larger than $p_{f,2}$ for $M_{s,2}$. The probability of failure p_f corresponds to the area under the PDF of M_s upper bounded by the vertical axis. From Figure 2.2 and Eq. (1.9), it can be seen that when the expected values of M_s are the same a larger uncertainty associated with M_s results in a larger probability of failure p_f (or smaller reliability p_s). Moreover, the structural performance indicators including reliability index β, redundancy index RI, vulnerability index VI, and robustness index RB, which are based on the probability of failure p_f or reliability p_s, are affected by the uncertainty associated with M_s.

2.2.2 Benefit from Accurate Bridge Life-Cycle Performance and Service Life Prediction

The accuracy of bridge performance and service life prediction can affect the life-cycle cost of a deteriorating bridge structure. Figure 2.3 shows the possible effects of uncertainties related to the initial structural performance P_{in} on the structural performance and service life prediction, where three cases (i.e., cases 1, 2 and 3) are considered.

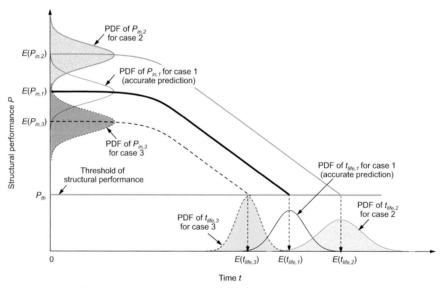

Figure 2.3 Effect of uncertainty related to initial structural performance on service life prediction.

- Case 1 represents the accurate prediction of structural performance P and service life t_{life}.
- Case 2 represents the nonconservative prediction of structural performance P and service life t_{life} for $E(P_{in,2}) > E(P_{in,1})$.
- Case 3 represents the conservative prediction of structural performance P and service life t_{life} for $E(P_{in,3}) < E(P_{in,1})$.

where $E(P_{in,i})$ is the initial structural performance of the ith case. Cases 2 and 3 are both inaccurate. It should be noted that the dispersion of the initial structural performance P_{in} is assumed the same for all the three cases considered. Compared to case 1, case 2 leads to later maintenance actions and unexpected increase in the probability of failure, and case 3 results in unnecessary maintenance actions. From Figure 2.3, it can be found that there is a benefit from accurate structural performance and service life prediction by applying the timely maintenance actions and reducing the probability of failure and expected failure cost. Therefore, it is important to use the structural performance prediction model with less uncertainty, by reducing the epistemic uncertainties as much as possible, for accurate and reliable bridge performance and service life management. If the information obtained from the inspection and monitoring is integrated with the existing structural performance prediction in a rational way, the uncertainties associated with the structural performance prediction can be significantly reduced (Estes and Frangopol 2003; Soliman and Frangopol 2014; Dong and Frangopol 2016a).

2.2.3 Loss Function for Structural Performance Prediction Error

The monetary loss can result from untimely and/or inappropriate maintenance actions and unexpected increase in the probability of failure of deteriorating bridges (Kim and Frangopol 2009; Frangopol and Kim 2014c). In order to quantify the monetary loss, the loss function based on the error between the actual and predicted parameters for the structural performance can be used. Figure 2.4 shows the loss functions f_{loss} based on constant, linear and quadratic forms, which are expressed, respectively, as (Ang and Tang 1984)

$$f_{loss}(x,x') = C_{loss} \qquad \text{for} \qquad e_{all} \leq |x - x'| \qquad (2.1a)$$

$$f_{loss}(x,x') = C_{loss} \cdot |x - x'| \qquad \text{for} \qquad -\infty \leq x \leq \infty \qquad (2.1b)$$

$$f_{loss}(x,x') = C_{loss} \cdot (x - x')^2 \qquad \text{for} \qquad -\infty \leq x \leq \infty \qquad (2.1c)$$

where x and x' are the actual and predicted parameters, respectively, C_{loss} is the unit monetary loss, and e_{all} is the allowable error. The expected monetary loss E_{loss} for the quadratic loss function f_{loss} is formulated as

$$E_{loss} = \int_{-\infty}^{\infty} f_{loss}(x,x') \cdot f_X(x)dx = C_{loss} \cdot \left\{(\sigma_x)^2 + (\mu_x - x')^2\right\} \qquad (2.2)$$

where $f_X(x)$ is the PDF of the parameter X associated with the structural performance, and σ_x and μ_x are the standard deviation (SD) and mean of X, respectively. Eq. (2.2) indicates that the expected loss E_{loss} increases with increases in the SD σ_x and the error between the mean of the parameter and predicted parameter (i.e., $\mu_x - x'$). Considering that accuracy of the predicted parameters can be improved by adding more samples associated with the parameter prediction, the quadratic loss function f_{loss} of Eq. (2.1c) can be modified as (Ang and Tang 1984)

$$f_{loss}(x,x') = C_{loss} \cdot (x-x')^2 + C_{spl} \cdot n_{spl} \tag{2.3}$$

where C_{spl} is the unit cost for sampling, and n_{spl} is the number of samples. Based on the minimum expected loss, the optimum sample size for inspection and SHM planning can be determined (Frangopol and Kim 2014b).

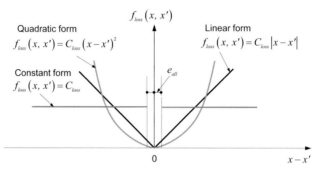

Figure 2.4   Constant, linear and quadratic loss functions.

2.2.4   Updating Bridge Life-Cycle Performance and Service Life Prediction

The new information obtained from each inspection and/or SHM can be integrated with the existing information on bridge performance and service life prediction to reduce the associated uncertainty (Frangopol 2011). The integration of the new information into the existing one can be achieved by using the Bayes' theorem-based updating techniques (Enright and Frangopol 1999a; Estes and Frangopol 2003; Zhu and Frangopol 2013b, 2013c). When the parameter Y is involved in the PDF $f_X(x)$ of a random variable X, and is represented by the PDF $f_Y'(y)$, the updated PDF $f_Y''(y)$ of the parameter Y is expressed using the Bayes' theorem as (Ang and Tang 2007)

$$f_Y''(y) = k \cdot L(y) \cdot f_Y'(y) \tag{2.4}$$

where k is the normalizing constant, and $L(y)$ is the likelihood function. The formulation of the likelihood function $L(y)$ is

$$L(y) = \prod_{i=1}^{n} f_X(x_i|y) \tag{2.5}$$

where n is the number of new data for the random variable X. Figure 2.5 illustrates the initial (i.e., prior), likelihood and updated PDFs (i.e., $f_Y'(y)$, $L(y)$ and $f_Y''(y)$, respectively). Based on the updated PDF $f_Y''(y)$, the PDF of the underlying random variable X can be obtained as

$$f_X''(x) = \int_{-\infty}^{\infty} f_X(x|y) \cdot f_Y''(y) \, dy \tag{2.6}$$

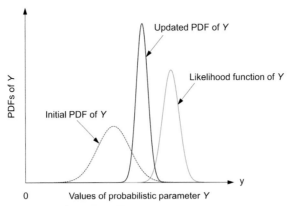

Figure 2.5 Initial (i.e., prior), likelihood and updated PDFs of probabilistic parameter *Y*.

When the underlying random variable X is normally distributed, the mean of X also follows a normal distribution, and n samples with the mean $\bar{x}$ and SD $\sigma/\sqrt{n}$ are available, the updated PDF $f''(\mu)$ of the mean of X becomes (Ang and Tang 2007)

$$f''(\mu) = k \cdot N_u\left(\bar{x}, \frac{\sigma}{\sqrt{n}}\right) \cdot N_u(\mu', \sigma') \qquad (2.7)$$

where μ' and σ' are the prior mean and SD of X, respectively, and N_u denotes the normal distribution of μ. The mean μ'' and SD σ'' of the updated PDF $f''(\mu)$ are given as

$$\mu'' = \frac{\bar{x}(\sigma')^2 + \mu'(\sigma^2/n)}{(\sigma')^2 + (\sigma^2/n)} \qquad (2.8a)$$

$$\sigma'' = \sqrt{\frac{(\sigma')^2(\sigma^2/n)}{(\sigma')^2 + (\sigma^2/n)}} \qquad (2.8b)$$

When the two parameters Y_1 and Y_2 involved in the PDF $f_X(x)$ of the underlying random variable X are updated simultaneously, the updated joint PDF $f_Y''(y_1, y_2)$ of Y_1 and Y_2, likelihood function $L(y_1, y_2)$, and the updated PDF $f_X''(x)$ are (Zhu and Frangopol 2013b, 2013c)

$$f_Y''(y_1, y_2) = k \cdot L(y_1, y_2) \cdot f_Y'(y_1, y_2) \qquad (2.9a)$$

$$L(y_1, y_2) = \prod_{i=1}^{n} f_X(x_i | y_1, y_2) \qquad (2.9b)$$

$$f_X''(x) = \int_{-\infty}^{\infty} \int_{-\infty}^{\infty} f_X(x | y_1, y_2) \cdot f_Y''(y_1, y_2) \, dy_1 dy_2 \qquad (2.9c)$$

Moreover, the non-conjugate distribution requires special techniques to use the Bayes' theorem, since the non-conjugate distribution cannot result in an explicit posterior distribution (Okasha and Frangopol 2012). Therefore, simulation techniques such as the Markov chain Monte Carlo (MCMC) sampling can be applied to generate efficiently the samples associated with a posterior PDF without determining the normalized constant k. The simulation of the Markov chain can be based on the cascaded Metropolis-Hastings and the slice sampling algorithms. Descriptions of these two algorithms are provided in Metropolis et al. (1953), Hasting (1970), Robert (1994), Chib and Greenberg (1995), Gilks et al. (1996), Tierney and Mira (1999), Neal (2003), Robert and Cassella (1999), and Rastogi et al. (2017).

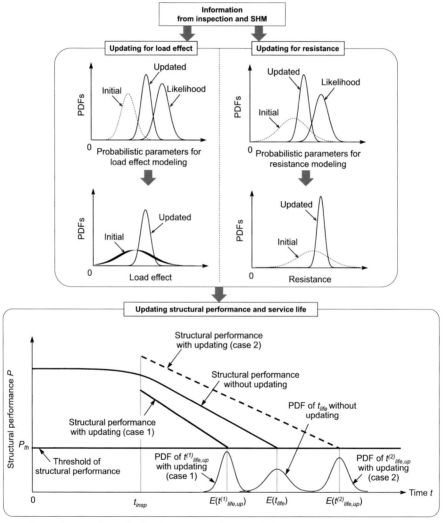

Figure 2.6   Schematic for the process from updating the parameters of PDFs to updating structural performance and service life.

Figure 2.6 shows the schematic for the process from updating the parameters of PDFs to updating structural performance and service life. Using the Bayes' theorem with the information from inspection and SHM, the single or multiple parameters of the PDFs representing the load effect and resistance can be updated. Finally, the bridge life-cycle performance and service life with the updated load effect and resistance can be re-estimated. Depending on the updated parameters associated with the load effect and resistance, the updated structural performance P_{up} and service life $t_{life,up}$ can be larger or less than the initially predicted one. Case 1 in Figure 2.6 corresponds to the case where the updated structural performance is less than initially predicted (i.e., $P_{up} < P$) and therefore, the updated expected service life is less than that predicted initially (i.e., $E(t^{(1)}_{life,up}) < E(t_{life})$). Case 2 in Figure 2.6 indicates $P_{up} > P$ and $E(t^{(2)}_{life,up}) > E(t_{life})$. The probabilistic approaches to use the monitoring data for updating single and multiple parameters were presented by Enright and Frangopol (1999a), Estes and Frangopol (2003), Estes et al. (2004), Okasha et al. (2010, 2012), Okasha and Frangopol (2012), Zhu and Frangopol (2013b, 2013c), Dong and Frangopol (2016a), and Thanapol et al. (2016), where the effects of updating the resistance and load effect on structural performance were investigated. Furthermore, Orcesi and Frangopol (2011a), Soliman and Frangopol (2014), Rastogi et al. (2017), Yang and Frangopol (2018c), and Kim et al. (2019) investigated the optimum inspection and maintenance planning for deteriorating structures considering the updating process with inspection and monitoring information.

2.3   BRIDGE PERFORMANCE DETERIORATION

Bridge performance continuously deteriorates over time under external loadings and environmental conditions. The continuous deterioration process of existing bridges is caused by the reduction of the resistance and/or increase in load effects. In general, the fatigue in steel bridges and corrosion in reinforced concrete bridges have been treated as the main deterioration mechanisms reducing bridge safety over time (Fisher 1984; Fisher et al. 1998; FHWA 2013b; NCHRP 2005, 2006). In order to predict the bridge performance deterioration, appropriate modeling of the fatigue and/or corrosion damage propagation is required (Kim et al. 2013; Frangopol and Kim 2019). Assessment and prediction of the load effect induced by traffic on bridges are also essential for bridge performance prediction. For live load modeling, occurrence and magnitude of live load should be considered using appropriate probabilistic approaches (Nowak 1999; NCHRP 2004).

2.3.1   Fatigue Damage Prediction

Repeated loading can lead to damage at fatigue-sensitive locations such as mechanically fastened connections with welding, rivets and bolts of steel structures (Haghani et al. 2012; Alencar et al. 2019). The fatigue damage initiation and propagation are affected by loading conditions, geometry and material properties of the fatigue-sensitive details, and initial defects (Fisher 1984; Li et al. 2016). The fatigue damage can be expressed in terms of the cumulative number of cycles and fatigue crack size (Hashin and Rotem 1978; Hwang and Han 1986).

Based on the cumulative number of cycles, the fatigue damage can be computed as (Miner 1945)

$$D_f = \sum_{i=1}^{n} \frac{n_i}{N_i} = \frac{N}{A} \cdot S_{re}^m \tag{2.10}$$

where D_f is Miner's damage accumulation index, which ranges from zero (i.e., no damage) to one (i.e., full damage); n_i is the number of cycles at the ith stress range level; N_i is the total number of cycles to failure at the ith stress range level; N is the total accumulated number of cycles; A is the fatigue detail coefficient; and m is the material constant. Based on the data collected from experiments and monitoring systems, the equivalent stress range S_{re} is computed as (Kwon and Frangopol 2010)

$$S_{re} = \left[\int_0^\infty s^m \cdot f_S(s) \, ds \right]^{1/m} = \left(\sum_{i=1}^{k} \frac{n_i}{N_{total}} \cdot S_{re,i}^m \right)^{1/m} \tag{2.11}$$

where $f_S(s)$ = PDF of the stress range S; n_i = number of observations for the stress range bin $S_{re,i}$; and N_{total} = total number of observations up to the kth stress range. The S-N curve approaches, which have been adopted by design standards and specification of steel structures (e.g., AASHTO (2002)), use the relation between S_{re} and N_{total}.

For fatigue crack damage prediction, Paris's law has been widely used among several empirical and phenomenological-based crack propagation models (Schijve 2003). Based on the Paris' law, the ratio of the crack size increment to number of cycles is defined as (Paris and Erdogan 1963)

$$\frac{da}{dN} = C(\Delta K)^m \quad \text{for} \quad \Delta K > \Delta K_{th} \tag{2.12}$$

where a is the crack size, N is the cumulative number of cycles, C and m are the material crack propagation parameters, ΔK is the stress intensity factor, and ΔK_{th} is the threshold of stress intensity factor. ΔK is expressed as (Irwin 1958)

$$\Delta K = S_{re} \cdot Y(a) \cdot \sqrt{\pi a} \tag{2.13}$$

where $Y(a)$ is the geometry function for a specific fatigue detail. Based on Eqs. (2.12) and (2.13), the cumulative number of cycles N is given as (Fisher 1984; Chung et al. 2006)

$$N = \frac{1}{C \cdot S_{re}^m} \int_{a_0}^{a_N} \left[Y(a)\sqrt{\pi a} \right]^{-m} da \tag{2.14}$$

where a_0 is the initial crack size, and a_N is the crack size after the N cumulative number of cycles. If the annual number of cycles N_{an} increases with the rate r_{an}, the time (years) t to reach the crack size a_N can be estimated as (Madsen et al. 1985, 1991; Frangopol and Kim 2019)

$$t = \frac{\ln\left[1 + \dfrac{\ln(1+r_{an})}{N_{an} \cdot C \cdot S_{re}{}^{m}} \cdot \displaystyle\int_{a_0}^{a_N} \left[Y(a)\sqrt{\pi \cdot a}\right]^{-m} da\right]}{\ln(1+r_{an})} \qquad \text{for } r_{an} > 0 \qquad (2.15a)$$

$$t = \frac{1}{N_{an} \cdot C \cdot S_{re}{}^{m}} \cdot \int_{a_0}^{a_N} \left[Y(a)\sqrt{\pi \cdot a}\right]^{-m} da \qquad \text{for } r_{an} = 0 \qquad (2.15b)$$

For the geometry function with a constant value (i.e., $Y(a) = Y$) and $r_{an} = 0$ in Eq. (2.15), the time t for the crack to propagate from a_0 to a_N is computed as (Guedes Soares and Garbatov 1996a, 1996b)

$$t = \frac{a_N{}^{(2-m)/2} - a_0{}^{(2-m)/2}}{\left(\dfrac{2-m}{2}\right) \cdot C \cdot S_{re}{}^{m} \cdot Y^{m} \cdot \pi^{m/2} \cdot N_{an}} \qquad \text{for } m \neq 2 \qquad (2.16a)$$

$$t = \frac{\ln(a_N) - \ln(a_0)}{C \cdot S_{re}{}^{m} \cdot Y^{m} \cdot \pi \cdot N_{an}} \qquad \text{for } m = 2 \qquad (2.16b)$$

Furthermore, the crack size a_N associated with time t of Eq. (2.15) is predicted as (Madsen et al. 1985; Chung et al. 2006; Frangopol and Kim 2019)

$$a_N = \left[a_0{}^{(2-m)/2} + \left(\frac{2-m}{2}\right) \cdot C \cdot S_{re}{}^{m} \cdot Y^{m} \cdot \pi^{m/2} \cdot N_{cy}\right]^{\left(\frac{2}{2-m}\right)} \qquad \text{for } m \neq 2 \qquad (2.17a)$$

$$a_N = a_0 \cdot \exp[C \cdot S_{re}{}^{m} \cdot Y^{m} \cdot \pi \cdot N_{cy}] \qquad \text{for } m = 2 \qquad (2.17b)$$

When the uncertainties associated with the variables of the fatigue crack propagation [e.g., a_0, S_{re}, N_{an} in Eq. (2.15)] are considered using Monte Carlo simulation and Latin Hypercube sampling methods, the time t and crack size a_N can be represented by PDFs. Figure 2.7 shows the fatigue crack propagation over time. The PDF of the crack size at a specific time is illustrated in Figure 2.7(a). Figure 2.7(b) shows the PDF of the time to reach a specific crack size. These PDFs can be used to predict the probabilistic structural performance indicators such as probability of failure p_f, reliability index β and structural redundancy index RI.

For fatigue reliability analysis based on the AASHTO Specification (AASHTO 2002) and Miner's rule (Miner 1945), the state function g is expressed as (Kwon and Frangopol 2011)

$$g(t) = \Delta - D_f(t) \qquad (2.18)$$

where Δ is the Miner's critical damage index, and D_f is the Miner's damage accumulation index defined in Eq. (2.10). The Miner's damage accumulation index D_f increases with an increase in the cumulative total number of cycles N over time as shown in Eq. (2.10). Considering the uncertainties associated with the variables Δ, N, A and S_{re}, the time-dependent reliability p_s [see Eq. (1.4a)], probability of failure p_f [see Eq. (1.4b)] and reliability index β [see Eq. (1.10)] can be computed with the state function of Eq. (2.18). Furthermore, the state function based on the crack size at time t can be formulated as (Kim and Frangopol 2011b; Dong and Frangopol 2016a)

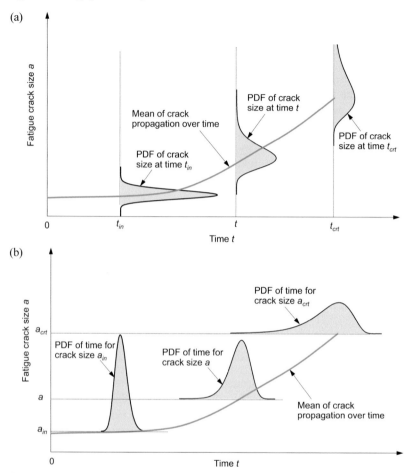

Figure 2.7   Fatigue crack propagation over time: (a) PDFs of crack size; (b) PDFs of time.

$$g(t) = a_{crt} - a(t) \tag{2.19}$$

where a_{crt} = critical crack size representing the maximum acceptable crack size threshold. The time-dependent crack size $a(t)$ is estimated using the random variables in Eq. (2.14), and can be represented by PDFs as shown in Figure 2.7(a). Using the PDFs of time to reach fatigue crack sizes as shown in Figure 2.7(b), the time-based state function at time t is given as (Kim and Frangopol 2011b)

$$g(t) = t_{crt} - t \tag{2.20}$$

where t_{crt} = time to reach the critical crack size a_{crt}. The general relation among time t, cumulative number of cycles N, crack size a and reliability index β is illustrated in Figure 2.8, where the cumulative number of cycles N increases over time, and accordingly, the fatigue crack size a increases, and finally the reliability index β decreases.

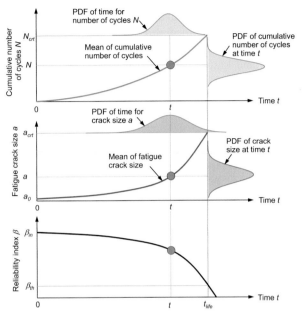

Figure 2.8 Relation among time, cumulative number of cycles, fatigue crack size and reliability index.

2.3.2 Corrosion Propagation Prediction

Corrosion in RC bridges affects the structural performance deterioration by reducing the cross sectional area and bond strength of the reinforcement in concrete (Arora et al. 1997; NCHRP 2005, 2006). The corrosion process in RC bridges is mainly caused by chloride penetration into concrete and concrete carbonation (Roberge 1999). In general, the corrosion deterioration process consists of six steps (Thoft-Christensen 2003): (a) penetration of chloride ions into the concrete structure; (b) corrosion initiation of reinforcement; (c) evolution of corrosion; (d) crack initiation of the concrete; (e) crack propagation; and (f) spalling. Based on Fick's second law, the chloride concentration $C(x, t)$ (g/mm^3) at time t is expressed as (Crank 1975)

$$\frac{\partial C(x,t)}{\partial t} = \frac{\partial}{\partial x}\left[D\frac{\partial C(x,t)}{\partial x}\right] \tag{2.21}$$

where x (mm) is the distance from the concrete surface, and D $(mm^2/year)$ is the effective chloride diffusion coefficient. Assuming that the effective chloride diffusion coefficient D and chloride concentration at surface C_0 are constant over time, and $C = 0$ at $t = 0$, the chloride concentration $C(x, t)$ in Eq. (2.21) is formulated as (Crank 1975)

$$C(x,t) = C_0 \left[1 - erf\left(\frac{x}{2\sqrt{D \cdot t}} \right) \right] \tag{2.22}$$

where erf = standard error function. When the chloride concentration C of reinforcement exceeds a predefined threshold C_{th}, corrosion of reinforcement initiates and propagates. Accordingly, the corrosion initiation time t_{cor} (years) is predicted as (Zhang and Lounis 2006)

$$t_{cor} = \frac{x^2}{4D \left[erfc^{-1}\left(\dfrac{C_{th}}{C_0} \right) \right]^2} \tag{2.23}$$

where $erfc$ is the complementary error function. As indicated in Eq. (2.23), the corrosion initiation can be delayed by increasing the distance from the concrete surface x, and/or reducing the effective chloride diffusion coefficient D and/or chloride concentration at surface C_0. Investigations about the probabilistic effects of x, C_{th}, D, and C_0 on the corrosion initiation time t_{cor} are available in Enright and Frangopol (1998a).

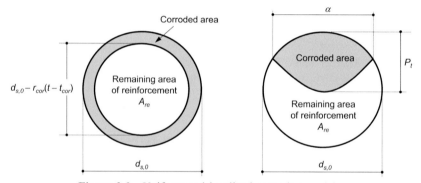

Figure 2.9   Uniform and localized corrosion models.

After the corrosion initiation time t_{cor}, reinforcement area can be reduced. The uniform and localized corrosion models have been widely adopted to predict the corrosion propagation and remaining area of the reinforcement (Val and Melchers 1997; Stewart 2004; Kim et al. 2013). Figure 2.9 illustrates the uniform and localized corrosion propagations. The uniform corrosion model assumes that the entire cross sectional area of reinforcement is reduced uniformly. Accordingly, the remaining area of the reinforcement A_{re} at time t is computed as (Enright and Frangopol 1998a; Marsh and Frangopol 2008)

$$A_{re}(t) = \frac{\pi d_{s,0}^{\,2}}{4} \qquad\qquad \text{for } 0 \le t \le t_{cor} \tag{2.24a}$$

$$A_{re}(t) = \frac{\pi}{4}\left[d_{s,0} - r_{cor}(t - t_{cor}) \right]^2 \qquad\qquad \text{for } t > t_{cor} \tag{2.24b}$$

where $d_{s,0}$ = initial diameter of reinforcement (mm), and r_{cor} = corrosion rate (mm/ year). Moreover, the remaining area of the reinforcement A_{re} under the localized corrosion is expressed as (Val and Melchers 1997)

$$A_{re}(t) = \frac{\pi d_{s,0}^2}{4} - \frac{1}{2}\left[\theta_1\left(\frac{d_{s,0}}{2}\right)^2 - \alpha\left|\frac{d_{s,0}}{2} - \frac{P_t(t)^2}{d_{s,0}}\right|\right] - \frac{P_t(t)^2}{2}\left[\theta_2 - \frac{\alpha}{d_{s,0}}\right]$$

$$(2.25a)$$

$$\text{for} \quad P_t(t) \le \frac{d_{s,0}}{\sqrt{2}}$$

$$A_{re}(t) = \frac{1}{2}\left[\theta_1\left(\frac{d_{s,0}}{2}\right)^2 - \alpha\left|\frac{d_{s,0}}{2} - \frac{P_t(t)^2}{d_{s,0}}\right|\right] - \frac{P_t(t)^2}{2}\left[\theta_2 - \frac{\alpha}{d_{s,0}}\right]$$

$$(2.25b)$$

$$\text{for} \quad \frac{d_{s,0}}{\sqrt{2}} < P_t(t) \le d_{s,0}$$

$$A_{re}(t) = 0 \qquad \text{for} \qquad P_t(t) > d_{s,0} \qquad (2.25c)$$

where $d_{s,0}$ is the initial diameter of the reinforcement (mm). The maximum pit depth $P_t(t)$ is

$$P_t(t) = r_{cor} R(t - t_{cor}) \qquad \text{for} \qquad t \ge t_{cor} \qquad (2.26)$$

in which R = ratio of maximum pit depth to uniform pit depth. The variables θ_1, θ_2 and α are defined as

$$\theta_1 = 2\arcsin\left(\frac{\alpha}{d_{s,0}}\right) \qquad (2.27a)$$

$$\theta_2 = 2\arcsin\left(\frac{\alpha}{2P_t(t)}\right) \qquad (2.27b)$$

$$\alpha = 2P_t(t)\left[1 - \left(\frac{P_t(t)}{d_{s,0}}\right)^2\right]^{0.5} \qquad (2.27c)$$

Figure 2.10 compares the remaining reinforcement area A_{re} for the uniform and localized corrosion model, where the initial diameter of the reinforcement $d_{s,0}$ is 50 mm, and the $r_{cor}(t - t_{cor})$ ranges from 0 to 20. It can be seen that the remaining reinforcement area A_{re} decreases with the increase of the ratio R of maximum pit depth to uniform pit depth. When the variables involved in the prediction of A_{re} are treated as random, the PDF of the remaining area A_{re} at time t and PDF of time t for a specific remaining area A_{re} can be obtained as shown in Figures 2.11(a) and 2.11(b), respectively.

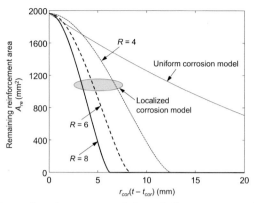

Figure 2.10   Comparison between remaining reinforcement areas for uniform and localized corrosion models.

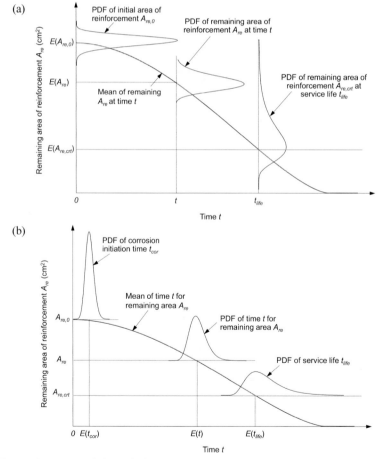

Figure 2.11   Remaining reinforcement area of RC bridges under corrosion: (a) PDFs of remaining reinforcement area for a given time; (b) PDFs of time associated with a given remaining reinforcement area.

The remaining reinforcement area A_{re} is used to predict the structural performance of RC bridges. For reliability analysis, the state functions in terms of the flexural and shear capacities can be formulated, respectively, as (Akgül and Frangopol 2005a)

$$g_{slab,flx} = M_R - M_L \qquad (2.28a)$$

$$g_{slab,shr} = V_R - V_L \qquad (2.28b)$$

where M_R = flexural capacity; M_L = moment by applied loads; V_R = shear strength; and V_L = shear force by applied loads. The definition of the variables M_R, M_L, V_R and V_L is available in AASHTO (2002) and Akgül and Frangopol (2005a and 2005b). M_R and V_R are affected by the remaining area of the reinforcement. Based on the state functions of Eq. (2.28), Akgül and Frangopol (2005a and 2005b) predicted the corrosion-based reliability index at component and system levels. Furthermore, the time-dependent reliability index of prestressed concrete and steel girder bridges under corrosion were investigated by Akgül and Frangopol (2004b) and (2004c), respectively.

2.3.3 Probabilistic Structural Performance Prediction

The generalized process for probabilistic structural performance prediction is presented in Figure 2.12. The time-dependent probabilistic structural performance indicators including probability of failure p_f and reliability index β can be computed with the state functions based on the fatigue and corrosion damage propagations over time. If the state function is formulated as a closed form, the probability of failure and reliability index can be obtained using several software programs as indicated in Section 1.3.2. If the probabilistic effects of inspection and maintenance on the structural performance are considered for optimum life-cycle cost of deteriorating bridges, it will be difficult to formulate the closed forms of the probabilistic performance indicators and state functions (Yang et al. 2006a; Barone and Frangopol 2014a). In this case, the simulation-based analysis methods such as Monte Carlo simulation and Latin Hypercube sampling should be used for computing the time-dependent probabilistic structural performance indicators.

When the time-dependent probabilistic performance indicators are used to optimize life-cycle performance and cost of deteriorating bridges, high computational time is generally required (Okasha and Frangopol 2010b). In order to improve the efficiency for computing probabilistic performance indicators and optimizing the life-cycle cost of deteriorating bridges, the lifetime distributions have been used extensively (Frangopol 2011; Frangopol and Soliman 2016). The detailed explanation on the lifetime distributions is provided in Section 1.3.3. The lifetime distributions of a structural component or system are formulated approximately with a predefined distribution types (e.g., exponential and Weibull distributions) and the associated parameters are obtained through the best fit of the analytical results and experimental data (Yang et al. 2006b). For this reason, it is difficult to find out the relation among the random variables involved in the damage propagation and load effects. Moreover, the structural performance based on the lifetime distribution is limited to the bridge system associated with only statistically independence or perfect correlation (Barone and Frangopol 2014a).

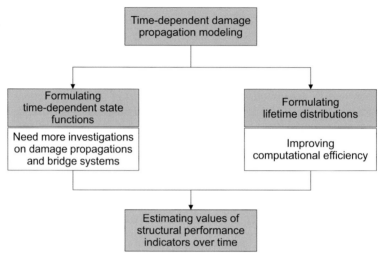

Figure 2.12   Schematic for the process for probabilistic structural performance prediction.

2.4   EFFECTS OF INSPECTION AND MAINTENANCE ON BRIDGE PERFORMANCE AND SERVICE LIFE

The structural performance continuously deteriorates over time as mentioned previously. In order to improve the structural performance and extend the service life, maintenance actions are applied (Sánchez-Silva et al. 2016). The appropriate maintenance types are determined based on the outcome of inspections (Kim et al. 2013; Frangopol and Kim 2019). The effects of inspection and maintenance on structural performance and service life considering the uncertainty associated with the damage propagation and inspection results are illustrated in Figure 2.13. When the inspection is performed at time t_{insp} and the damage is not detected or if the detected damage requires no maintenance, there will be no improvement of structural performance and no service life extension. According to the type of maintenance after damage detection, the improvement of structural performance and service life extension will vary. The application of preventive maintenance A in Figure 2.13 results in delay of structural performance deterioration and an increase in the mean initial service life $E(t_{life,0})$ provided by the difference $E(t_{life,A}) - E(t_{life,0})$. As indicated in the figure, $E(t_{life,0})$ is extended to the mean service life $E(t_{life,B})$ after the application of the essential maintenance B.

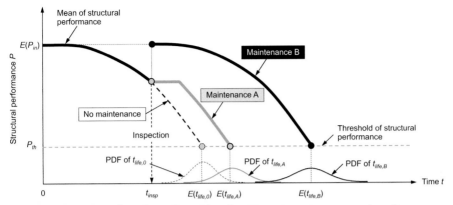

Figure 2.13 Effects of inspection and maintenance on mean structural performance and service life under uncertainty.

2.4.1 Formulation of Extended Service Life

The service life considering inspections and maintenances can be formulated using the event tree shown in Figure 2.14. This event tree is based on the structural performance profiles with a single inspection at time $t_{insp,1}$ and two maintenance types (i.e., Maintenances A and B) shown in Figure 2.13. The event tree represents all the possible outcomes and associated occurrence probabilities. The outcome of the event tree is the extended service life. The occurrence probabilities consist of the probabilities that the inspection is performed within the service life, damage is detected by the inspection, and the maintenance is applied. These probabilities are caused by the uncertainties associated with the damage propagation, service life prediction, and damage detection. As shown in Figure 2.14, if the inspection is applied before the initial service life $t_{life,0}$, and damage is detected, an in-depth inspection will be performed to identify the degree of damage. Maintenance A is applied when the degree of damage is identified to be between d_A and d_B (see Branch 2). For the identified damage larger than d_B, Maintenance B will be applied (see Branch 3). No maintenance results from three cases as: (a) the detected degree of damage is less than a predefined limit d_A (see Branch 1); (b) damage is not detected (see Branch 4); and (c) inspection is applied later than the initial service life $t_{life,0}$ (see Branch 5).

The formulation of the expected extended service life $t_{life,ex}$ is

$$t_{life,ex} = \sum_{i=1}^{n_b} t_{life,i} \cdot P_{b,i} \qquad (2.29)$$

where n_b = number of branches in the event tree; $t_{life,i}$ = extended service life for the ith branch; and $P_{b,i}$ = occurrence probability of the ith branch. For example, the expected extended service life $t_{life,ex}$ associated with the event tree of Figure 2.14 is formulated as

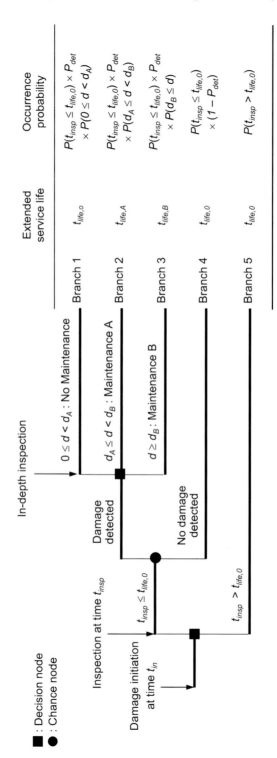

Figure 2.14 Event tree to formulate the expected extended service life under uncertainty.

$$t_{life,ex} = t_{life,0} \times P(t_{insp} \leq t_{life,0}) \times P_{det} \times P(0 \leq d < d_A)$$
$$+ t_{life,A} \times P(t_{insp} \leq t_{life,0}) \times P_{det} \times P(d_A \leq d < d_B)$$
$$+ t_{life,B} \times P(t_{insp} \leq t_{life,0}) \times P_{det} \times P(d_B \leq d) \qquad (2.30)$$
$$+ t_{life,0} \times P(t_{insp} \leq t_{life,0}) \times (1 - P_{det}) + t_{life,0} \times P(t_{insp} > t_{life,0})$$

where $t_{life,A}$ and $t_{life,B}$ are the extended service life by applying Maintenances A and B, respectively. The probability of damage detection is denoted as P_{det} in Eq. (2.30). As shown in Figure 2.14, the outcome of Branch 1 is the initial service life $t_{life,0}$, since there is no maintenance. The occurrence probability associated with Branch 1 considers three events: (a) inspection is performed before the initial service life (i.e., $t_{insp} \leq t_{life,0}$); (b) inspection detects the damage; and (c) degree of the detected damage ranges between 0 and d_A. Accordingly, the occurrence probability becomes $P(t_{insp} \leq t_{life,0}) \times P_{det} \times P(0 \leq d < d_A)$. In this manner, the extended service life and occurrence probability of each branch can be obtained as indicated in Figure 2.14, and the expected extended service life $t_{life,ex}$ is formulated as shown in Eq. (2.30).

2.4.2 Effects of Inspection Quality and Time on Extended Service Life

In order to investigate the effects of inspection quality and time on the extended service life, suppose that a bridge is deteriorating over time, and its time-dependent damage intensity $d(t)$ is expressed as (Kim et al. 2013)

$$d(t) = 0 \qquad\qquad \text{for} \qquad 0 \leq t \leq t_{cor} \qquad (2.31a)$$

$$d(t) = \exp[(t - t_{cor})/\lambda] - 1 \qquad \text{for} \qquad t > t_{cor} \qquad (2.31b)$$

where $d(t)$ is the damage intensity at time t. The damage initiation time t_{cor} follows a lognormal distribution with a mean of 5 years and a SD of 1 year (denoted as LN(5 years; 1 year)). The scale parameter λ in Eq. (2.31b) is assumed lognormally distributed with LN(50 years; 10 years). This example assumes that the initial service life $t_{life,0}$ is the time for the damage intensity $d(t)$ to reach 0.5, and two types of maintenance (i.e., Maintenances A and B) are applied to extend the service life. Maintenances A and B result in the service life extensions, which are represented by lognormal random variables LN(10 years; 2 years) and LN(20 years; 4 years), respectively. The damage intensity criteria d_A and d_B to determine the maintenance type [see Figure 2.14 and Eq. (2.30)] are assumed as 0.2 and 0.4, respectively. The probability of damage detection P_{det} is (Kim et al. 2013)

$$P_{det} = 1 - \Phi\left(\frac{\ln(d) - \ln(\alpha_d)}{\delta_d} \right) \qquad (2.32)$$

where Φ = standard normal cumulative distribution; and d = damage intensity defined in Eq. (2.31). α_d and δ_d indicate the location and scale parameters, respectively. The parameter δ_d is assumed to be $-0.1 \ln(\alpha_d)$. In this example, the parameter α_d is related to the quality of inspection. A smaller value of α_d results in higher P_{det}.

(a)

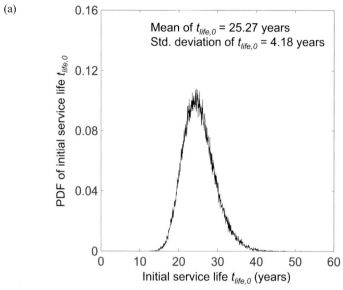

(b)

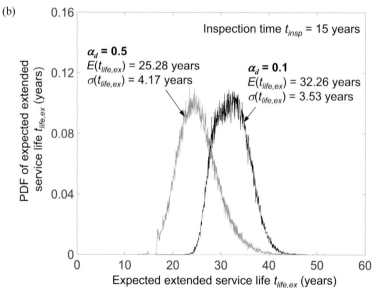

Figure 2.15   (a) PDF of initial service life; (b) PDF of the expected extended service life.

The PDFs of the initial service life $t_{life,0}$ and expected extended service life $t_{life,ex}$ are illustrated in Figure 2.15. The expected extended service life $t_{life,ex}$ is computed using Eq. (2.30) with the inspection time $t_{insp} = 15$ years. It should be noted that $t_{life,ex}$ is treated as a random variable, since the uncertainties associated with the initial service life and the extended service lives after Maintenances A and B are considered in this example. As shown in Figure 2.15(b), the mean of the expected extended service life $E(t_{life,ex})$ with $\alpha_d = 0.1$ is larger than $E(t_{life,ex})$ with $\alpha_d = 0.5$. Figure 2.16 shows the relation between the values of α_d and $E(t_{life,ex})$. It can be seen that a higher quality of inspection method (i.e., a smaller value of α_d) produces a larger $E(t_{life,ex})$. If the inspection time t_{insp} increases from 10 to 20 years for given α_d, $E(t_{life,ex})$ will increase. However, $E(t_{life,ex})$ for $t_{insp} = 25$ years is not always larger than $E(t_{life,ex})$ for $t_{insp} = 20$ years as shown in Figure 2.16. For this reason, in order to maximize $E(t_{life,ex})$, the optimum inspection time t_{insp} needs to be determined. The optimum inspection planning will be addressed in Chapter 6.

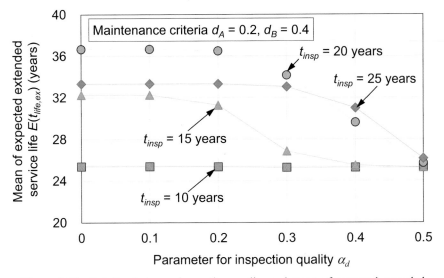

Figure 2.16 Relation between inspection quality and mean of expected extended service life for the inspection times t_{insp} = 10 years, 15 years, 20 years and 25 years.

2.4.3 Effects of Maintenance on Extended Service Life

The effect of maintenance criteria d_A and d_B in Eq. (2.30) on the mean of the expected extended service life $E(t_{life,ex})$ is investigated in Figure 2.17, when the inspection is performed at $t_{insp} = 15$ years. The assumptions used for this figure are the same as those used for Figures 2.15 and 2.16. Figure 2.17 considers three cases with the following degrees of damage as maintenance criteria: (a) case I: $d_A = 0.1$, $d_B = 0.3$; (b) case II: $d_A = 0.2$, $d_B = 0.4$; (c) case III: $d_A = 0.3$, $d_B = 0.5$. The degrees of damage d_A and d_B for case I are the smallest among these three cases. It means that the probability of maintenance for case I will be higher than those for cases II and III. Accordingly, when the maintenance criteria of case I is adopted, the mean of the expected extended service life $E(t_{life,ex})$ is higher than those associated with cases II and III. Furthermore, a reduction in inspection quality (i.e., larger value of α_d) leads to a reduction in the difference among the values of $E(t_{life,ex})$ for cases I, II and III as shown in Figure 2.17.

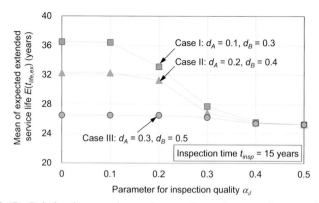

Figure 2.17   Relation between inspection quality and mean of expected extended service life for three cases with different maintenance criteria.

2.5   CONCLUDING REMARKS

This chapter deals with the uncertainties related to the structural performance and service life of deteriorating bridges. The cost-benefit resulting from the accurate bridge life-cycle performance and service life prediction is discussed. The loss function is presented in order to estimate the monetary loss resulting from untimely and/or inappropriate maintenance actions and unexpected increase in the probability of failure of deteriorating bridges. Since the new information from inspections and SHM can reduce the uncertainty in structural performance prediction, an appropriate updating process needs to be integrated in the life-cycle bridge performance. For this reason, the updating approaches developed recently and their applications are provided. Furthermore, the corrosion and fatigue deterioration processes for bridges are presented. The structural performance prediction is described by considering the state functions of the structural components and systems and lifetime distributions. It was shown that the structural performance and service life are affected by inspection quality and its time of application, and maintenance actions. The approaches presented herein for structural performance and service life prediction considering inspections and maintenances, can be used as part of the process of the optimum service life management of deteriorating bridges.

Chapter 3

Inspection and Structural Health Monitoring

NOTATIONS (continued)	
t_{ub}	= upper bound of damage occurrence time
S_{re}	= equivalent stress range
Δ	= Miner's critical damage index
$\eta(t)$	= time-dependent function at time t
$\sigma(t)$	= stress induced by live load at time t
$\sigma_0(t)$	= predefined stress limit at time t
σ_e	= standard deviation of measurement error e
σ_{max,N_t}	= largest predicted stress among the stresses induced by N_t heavy vehicles during a future time period t
σ_{mon}	= stress associated with the monitored strain

ABSTRACT

In Chapter 3, several representative inspection and SHM techniques for fatigue-sensitive steel bridges and RC bridges under corrosion are reviewed. The effect of inspection and SHM on life-cycle cost analysis is described. The probabilistic approaches to predict the bridge reliability using the data from monitoring and to quantify the availability of monitoring data are reviewed. The effects of monitoring and prediction durations on the expected availability of monitoring data are provided. The expected damage detection time and delay, and probability of failure are formulated considering the uncertainties associated with the damage occurrence, propagation, and detection. The relations among the probability of damage detection of an inspection method, monitoring duration, and number of inspections and monitoring, expected damage detection delay and probability of failure are investigated. Based on the probabilistic concepts presented in this chapter, the life-cycle inspection, monitoring, and maintenance management planning can be optimized.

3.1 INTRODUCTION

The main difficulties associated with the assessment of bridge safety and the optimization of service life management are generally related to the uncertainties associated with prediction of the damage occurrence and propagation, and bridge performance (Biondini and Frangopol 2016; Frangopol and Soliman 2016). In order to reduce the epistemic uncertainties in bridge management, and to prevent the unexpected unserviceability and structural failure of deteriorating bridges, inspection and structural health monitoring (SHM) should be applied in appropriate ways (NCHRP 2006; Verma et al. 2013; Copelan 2014). The inspection and SHM techniques have been applied for various purposes such as timely damage detection and identification, accurate condition assessment and prediction, and application of outcomes from inspection and SHM for effective and efficient bridge performance assessment and prediction (Okasha and Frangopol 2012;

Frangopol et al. 2008a, 2008b). Therefore, it is essential to understand the role of inspection and SHM in supporting the rational decision making for bridge safety and service life management.

This chapter presents representative inspection methods for fatigue-sensitive steel bridges and RC bridges under corrosion. The vibration-based, strain-based, vision-based, and GPS-based bridge SHM techniques are reviewed. The effect of inspection and SHM on life-cycle cost analysis is described. The data from inspection and monitoring can be used to predict structural performance under uncertainty. The probabilistic approaches to predict the bridge reliability using the data from monitoring and to quantify the availability of monitoring data are presented, where the monitoring data are produced from both the controlled and uncontrolled loading conditions. Considering the uncertainties associated with the damage occurrence, propagation, and detection, formulations of the damage detection time and delay for inspection and SHM applications are provided. Furthermore, an approach to estimate the probability of failure based on the damage detection time and delay is presented. The probabilistic approaches presented in this chapter can be used for optimum life-cycle inspection, monitoring, and maintenance management planning.

3.2 APPLICATIONS OF INSPECTION AND MONITORING FOR BRIDGE MANAGEMENT

The main purposes of the application of inspection and SHM for deteriorating bridges include (a) condition assessment, (b) damage detection and identification of critical locations, and (c) collection of data for accurate and reliable bridge performance and service life management (Frangopol et al. 2008a, 2008b). Over the last few decades, various inspection and SHM techniques, and extensive data processing methods have been developed to improve the efficiency and effectiveness of bridge management (Martinez-Luengo et al. 2016; Seo et al. 2016; Ostachowicz et al. 2019). Inspection and SHM techniques include non-destructive test, vibration-based, strain-based, vision-based, and GPS SHM systems (Seo et al. 2016; Xu and Brownjohn 2018). Along with the development of advanced inspection and SHM techniques, demands to integrate the information from inspection and SHM into structural performance assessment and prediction, and life-cycle management of deteriorating bridges are increasing (Frangopol and Soliman 2016).

3.2.1 Inspection Methods

Non-destructive inspection (NDI) is one of the most representative damage and condition assessment methods for deteriorating bridges. The inspection results depend on the inspection methods to be applied, inspectors' experiences and skills, and environmental conditions where the inspection is performed (Washer et al. 2014; Campbell et al. 2020). The essential requirements of the procedures and frequency of inspection, inspectors' qualifications, and documentation for the inspection results are recommended in FHWA (2004, 2012), and AASHTO (2017b). In general, bridge inspection needs to be applied at the uniform time interval of two years. Considering the current bridge condition and expected deterioration progress of bridge performance, the inspection can be applied more frequently (Copelan 2014).

Inspection methods for fatigue-sensitive steel bridges

For fatigue-sensitive steel structures, NDI techniques can detect the changes in material properties and cracks in steel details. The representative NDI techniques include the visual inspection (VI), ultrasonic inspection (UI), magnetic particle inspection (MPI), penetrant inspection (PI), radiographic inspection (RI), and acoustic emission inspection (AEI), among others (Hartle et al. 1995; Chung et al. 2006). The VI method is a primary inspection method used to evaluate the condition of deteriorating civil infrastructure such as bridges, buildings, and nuclear power plants. The VI method is generally performed to identify suspected cracks before applying an in-depth inspection using other NDI techniques (e.g., UI, MIP, PI, RI, and AEI) (Fisher et al. 1998). The accuracy and reliability of outcomes from the VI are highly affected by human factors (e.g., experience, education and training of inspectors), environmental factors (e.g., field temperature, wind speed, and lighting level), and organization and social factors (FHWA 2001; Washer et al. 2014; Campbell et al. 2020). In general conditions, the minimum detectable crack size of the VI method ranges from 3.0 mm to 6.0 mm (Fisher et al. 1998).

The UI method detects the cracks in steel bridge members using high-frequency waves. When the waves meet discontinuities and defects, distortion occurring in the waves is detected by the UI system (Fisher et al. 1998). This method can detect both internal and external cracks. The preparation for the UI is simple, and outcomes after inspection can be immediately obtained. However, the UI is affected by vibrations induced by live loads on bridges and requires considerable experience of inspectors (Shaw 2002; Martinez-Luengo et al. 2016).

In the MPI method, fine magnetic particles are sprayed on the inspecting locations. The distorted magnetic field is caused by the cracks in steel bridge members. The required equipment for the MPI is portable, and preparation before the inspection is simple. This method is effective to detect the cracks on smooth surfaces. However, it is difficult to detect cracks in welded connections and subsurfaces. The MPI method requires experienced inspectors (Fisher et al. 1998).

The PI method is one of the most widely used type of inspection because it is cost-effective and easy to use (ASTM 2012). When the liquid dye is applied to the surface to be inspected, and all of the dye on the surface is wiped clean, the dye penetrated into a crack cannot be removed, and remains in the crack. Finally, inspectors can recognize the crack on the surface (CDOT 2018). This inspection method requires removing an existing paint, and is limited to the detection of surface-breaking cracks (Fisher et al. 1998).

The RI method can detect cracks using x-rays or gamma rays. In general, radiation penetrating through a material is absorbed and scattered. Discontinuities and defects result in more radiation to pass than intact areas. The difference in the intensity of the radiation is visualized in the radiographic image, and finally, discontinuities and defects can be identified. Although the RI method is generally used during the fabrication of steel members and structures, it is still effective to assess inaccessible details and components of in-service bridges (Washer 2014).

The AEI uses a high frequency sound wave to detect failure mechanisms up to the microscale. This method allows a simple and cost-effective inspection of deteriorating steel bridges. Using AE-based sensors, the fatigue critical locations can be monitored continuously (Clemena et al. 1995). However, this method requires an appropriate filtering algorithm to eliminate external noise (Martinez-Luengo et al. 2016).

Inspection methods for RC bridges

Corrosion-induced cracking is one of the main damages affecting the structural performance of RC bridges (NCHRP 2006). The representative NDI techniques to assess the condition of RC bridges include the ultrasonic pulse velocity inspection (UPVI), impact echo inspection (IEI), AEI, and RI (IAEA 2002; Washer 2014). The UPVI can assess the condition of concrete by analyzing the amplitude and travel time of the acoustic wave. Equipments used for the UPVI are simple and portable. The required training to use this method is relatively easy compared to other NDI techniques. However, the UPVI needs an access to the both sides of a bridge member in order to obtain the acoustic waves from the transmitter to the receiver. This inspection may be time-consuming to investigate a large area, since this method is a point-based technique (Dilek 2007; Washer 2014).

The condition of RC bridges can be assessed by the IEI method, where the stress wave generated by impact is used. The primary advantage of this method is that it needs application to only one side of an RC bridge member, and therefore, this inspection can be applied to various types of RC members such as RC decks, concrete box girders and abutments. However, reliability of the outcomes from the IEI decreases when the thickness of an RC member increases. Also, experienced inspectors are required, since the accuracy of the inspection results depends on impact duration (Verma et al. 2013).

Although the AEI is traditionally used to detect the cracks in steel bridges, recent advanced techniques to apply these inspection methods for RC bridges subjected to corrosion have been extensively developed (Washer 2014). In general, the data from AEI for RC bridges need to be filtered using appropriate data processing algorithms, since the original data from AEI contain useless noises (Grosse 2010). Several investigations under controlled loading conditions reveal that the AEI can successfully detect and monitor crack initiation and propagation in concrete. However, application of the AEI to existing RC bridges under uncontrolled loadings should be based on the appropriate noise filtering algorithm, and be combined with the results from other NDI techniques to improve its accuracy and reliability (Ziehl et al. 2008; Sagar and Dutta 2021).

The RI can detect damages in reinforcements and prestressing strands, and voids in concrete (Washer 2014). When the radiation travels through RC bridge members, reinforcement bars, concrete, and voids in the concrete would absorb the different quantities of the radiation. Using the radiographic film, the sensitivity to radiation can be illustrated, and finally inspectors can detect corrosion in reinforcements, voids, and cracks in concrete. However, an increase in the thickness of the inspected concrete member requires more cost and time for inspection, higher quality image to represent damages in RC bridge members, and additional shielding system to prevent scattered radiation (IAEA 2002).

In order to determine the most appropriate inspection method for steel and RC bridges, types of damage to be inspected, accuracy of inspection results, accessibility to the inspected details, inspection cost, and probability of damage detection should be all considered in a rational way (Fisher et al. 1998; IAEA 2002; Copelan 2014).

Probability of damage detection

Uncertainty associated with an inspection method is unavoidable so that an existing damage cannot be detected with 100% probability. The quality of an inspection method can be represented by probability of damage detection, which is defined as the probability that the existing damage is detected by an inspection method (Frangopol and Kim 2019). In general, an increase in damage intensity results in a higher probability of damage detection. In order to quantify the relation between damage intensity and probability of damage detection, several forms of probability of damage detection P_{det} can be used such as:

(a) Normal CDF form (Frangopol et al. 1997b)

$$P_{det} = \Phi\left(\frac{d - d_{0.5}}{\sigma_d}\right) \tag{3.1}$$

where $\Phi(\cdot)$ = standard normal cumulative distribution function (CDF); d = damage intensity; $d_{0.5}$ = damage intensity associated with probability of damage detection equal to 0.5; and σ_d = parameter representing the uncertainty associated with $d_{0.5}$. The value of $d_{0.5}$ depends on the quality of an inspection method. A decrease in the value of $d_{0.5}$ indicates higher quality of inspection.

(b) Lognormal CDF form (Crawshaw and Chambers 1984)

$$P_{det} = 1 - \Phi\left(\frac{\ln(d) - \chi}{\gamma}\right) \tag{3.2}$$

where χ and γ are the location and scale parameters, respectively. The quality of an inspection method depends on these two parameters.

(c) Shifted exponential form (Packman et al. 1969):

$$P_{det} = 1 - \exp\left(-\frac{d - d_{min}}{\lambda}\right) \quad \text{for} \quad d > d_{min} \tag{3.3}$$

where d_{min} is the minimum detectable damage intensity, and λ is the scale parameter. For given damage intensity, a decrease in the value of d_{min} and/or λ leads to higher quality of inspection (or probability of damage detection). It should be noted that the scale parameter λ is larger than zero, and the damage intensity less than d_{min} is associated with $P_{det} = 0$.

(d) Log-logistic form (Berens and Hovey 1981; Berens 1989)

$$P_{det} = \frac{\exp[\rho + \kappa \ln(d)]}{1 + \exp[\rho + \kappa \ln(d)]} \tag{3.4}$$

where ρ and κ are the statistical parameters estimated using the maximum likelihood method. The expressions of P_{det} shown in Eqs. (3.1) to (3.4) can be used to formulate the objectives for optimum inspection and maintenance planning.

3.2.2   Structural Health Monitoring Techniques

SHM systems continuously measure the changes in structural characteristics and behaviors of deteriorating bridges. The changes beyond a predefined threshold are treated as damage through damage detection algorithms (Farrar and Worden 2007). The structural characteristics and behaviors of bridges can be continuously monitored by measuring vibrations, strains, and local and global displacements, among others (Doebling et al. 1998; Seo et al. 2016; Ostachowicz et al. 2019).

Vibration-based SHM systems

Vibration-based SHM systems have been commonly used in deteriorating bridges over the last few decades (Peeters et al. 2001; Ostachowicz et al. 2019). The vibration-based SHM systems measure the dynamic characteristics such as natural frequency, modal strain energy, mode shape, and dynamic flexibility, among others (DeWolf et al. 2002; Seo et al. 2016). The associated techniques are generally based on the premise that damage will produce changes in the dynamic characteristics of a bridge member or system. Among the dynamic characteristics, the natural frequencies of the bridge components and system have been used to investigate the global and local structural damage. Using probabilistic and statistical algorithms, the damage with its location and magnitude can be identified. The modal strain energy can be also used to detect and localize the damage. The strain energy may be more sensitive to damage than natural frequency (Doebling et al. 1997; Seo et al. 2016). The mode shape curvature can allow localized damage detection, where the stiffness changes of structural members are associated with damaged structural members (Lee et al. 2007). Dynamic flexibility has been adopted to detect damage in bridges, where the change in flexibility matrix before and after damage occurrence is analyzed (Ta et al. 2006; Pandey and Biswas 1995). Additional information about vibration-based SHM is available in Doebling et al. (1998), Peeters et al. (2001), DeWolf et al. (2002), Carden and Fanning (2004), and Brownjohn et al. (2011).

Strain-based SHM systems

Strain-based SHM systems, which collect the strain data from controlled and uncontrolled loading conditions on the monitored bridge, can be used to identify the fatigue damage of steel bridges (Kwon and Frangopol 2010, 2011). The controlled loading tests include static park test, crawl test associated with a low speed truck (e.g., up to 10 km/h), and dynamic test associated with a high speed truck (e.g., up to 80 km/h) (Connor and Santosuosso 2002; Connor and McCarthy 2006). With the monitored data from the controlled loading conditions, the maximum stresses and strains can be determined, and the analytical investigations based on finite element models of bridges can be verified. This information can be used to assess the structural performance of deteriorating bridges (Frangopol et al. 2008a, 2008b; Orcesi and Frangopol 2010). Figure 3.1 shows the monitored stress-time histories associated with the controlled test trucks passing over the right and left lanes, where the strain gages were installed at the bottom face of

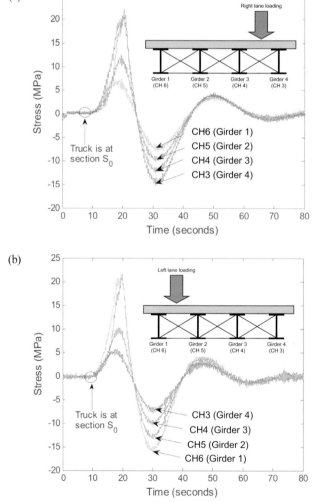

Figure 3.1 Monitored stress-time histories for controlled crawl test: (a) test truck over right lane; (b) test truck over left lane (adapted from Orcesi and Frangopol 2010).

the bottom flange of each girder, and the monitored strain is converted into stress using Hooke's law (Mahmoud et al. 2005). Orcesi and Frangopol (2010) used the results of the crawl tests (see Figure 3.1) to estimate the influence lines for bending moment at sensor locations. From these influence lines, the maximum bending moment produced by trucks and the reliability in terms of serviceability are computed.

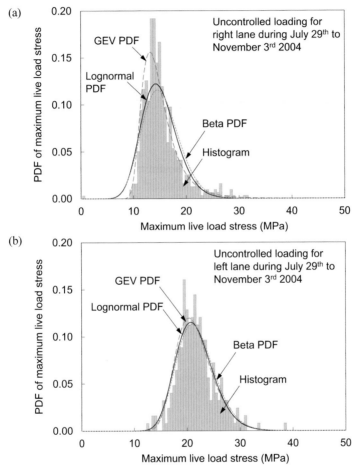

Figure 3.2 Histogram and PDFs of maximum live load stress obtained from uncontrolled loading: (a) right lane loading; (b) left lane loading (adapted from Liu et al. 2009b).

The monitored data under an uncontrolled loading (i.e., actual traffic) condition can provide the maximum live load stresses during a predefined time interval, and determine the most appropriate probability density function (PDF) associated with the maximum live load stress. This PDF is used to assess and predict the reliabilities of bridge components and system (Liu et al. 2009a, 2009b). Figure 3.2 shows the histogram obtained from the monitored live load stresses, which were collected from the uncontrolled loadings, and the several PDFs (e.g., lognormal, generalized extreme values (GEV) and beta PDFs) to represent the histogram. Detailed information on Figure 3.2 is available in Liu et al. (2009b). Furthermore, the monitored data under uncontrolled loadings can be used for fatigue reliability assessment and prediction. Based on the S-N curves presented in the AASHTO design specification (2010), the state function [see Eq. (2.18)] is formulated, where the stress range and number of cycles obtained from the monitored data serve as input random variables (Liu et al. 2010a; Kwon and Frangopol 2010, 2011). The monitored live load stress can also be integrated into an existing finite element model and life-cycle performance prediction model of the monitored bridge through the updating process (Okasha et al. 2010, 2012).

Vision-based SHM systems

Vision-based SHM systems allow monitoring local and global displacements and external conditions of in-service bridges (Xu and Brownjohn 2018). Vibration and strain-based SHM systems have several drawbacks related to efforts and costs for SHM system installation, operation and management. Along with the developments in high-resolution cameras, optical sensors and digital image processing techniques, the vision-based SHM systems have been investigated and applied for bridge management. The vision-based systems allow easy installation, non-contact and remote monitoring, and multi-point measurements using a single optical instrument for a large structure (Jahanshahi et al. 2009; Myung et al. 2014; Feng and Feng 2015). However, the results of this SHM system for damage detection are affected by weather conditions, structural vibration, image quality and image-processing techniques (Fukuda et al. 2010; Feng and Feng 2018). In order to improve the accessibility to critical locations of an inspected bridge, and efficiency to collect images of multiple locations, the vision-based SHM systems using the unmanned aerial vehicles (UAV) have been recently investigated (Akbar et al. 2019). However, more investigations are still required to address further improvements related to battery life, carrying capacity, wind-induced vibration control, distance and cost of data transfer systems (Morgenthal and Hallermann 2014; Ellenberg et al. 2015; Yoon et al. 2018).

GPS SHM systems

Global positioning system (GPS) is able to continuously measure global displacements of existing bridges under all weather conditions where the static displacement and dynamic characteristics can be assessed simultaneously (Casciati and Fuggini 2009). This SHM system is suitable to monitor displacements of structures with relatively low natural frequency such as high-rise buildings, long-span bridges, large-scale dams, towers and chimneys. The measurement errors of GPS are caused by multipath effects, geometry effects of satellites, tropospheric delay, and cycle slip, among others (Im et al. 2013). Depending on the monitored structures, GPS sensors, and data processing algorithms to be adopted, the recent GPS SHM system can measure the vertical and horizontal displacement of less than 3.0 mm and 1.0 mm, respectively (Nickitopoulou et al. 2006; Casciati and Fuggini 2009; Yi et al. 2010; Im et al. 2013). The accuracy of the GPS system can be significantly improved by integrating with other types of SHM systems including vibration-based, strain-based and vision-based SHMs (Myung et al. 2014).

3.2.3 Effect of SHM on Life-Cycle Cost Analysis

The application of inspection and SHM to deteriorating bridges can lead to more accurate and reliable structural performance and service life prediction, and finally affect the total life-cycle cost. Based on Eq. (1.1), the total life-cycle cost $C_{life,m}$ considering SHM can be expressed as (Frangopol and Messervey 2011)

$$C_{life,m} = C_{int} + C_{ins,m} + C_{ma,m} + C_{fail,m} + C_{mon} \tag{3.5}$$

where C_{int} = initial cost; $C_{ins,m}$ = scheduled inspection cost; $C_{ma,m}$ = cost for maintenance and repair actions; $C_{fail,m}$ = expected failure cost; and C_{mon} = cost for SHM. The subscript m indicates the costs with SHM (i.e., $C_{ins,m}$, $C_{ma,m}$ and $C_{fail,m}$). These costs are affected by SHM, because estimations of $C_{ins,m}$, $C_{ma,m}$ and $C_{fail,m}$ depend on the epistemic uncertainties associated with structural performance and service life prediction. The SHM can reduce these uncertainties. As described in Section 2.2.2, the benefit from the accurate life-cycle performance and service life prediction can be obtained through the timely maintenance actions and reduction in the probability of failure and expected failure cost. The cost-benefit of SHM $C_{ben,m}$ can be quantified by estimating the difference between the total life-cycle costs with and without SHM (i.e., $C_{ben,m} = C_{life} - C_{life,m}$) (Frangopol and Kim 2019).

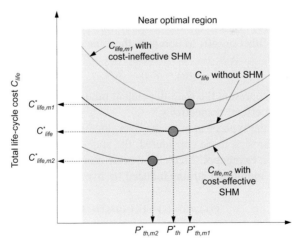

Figure 3.3 Relation between structural performance threshold P_{th} and total life-cycle cost C_{life} of an existing bridge when the SHM is not applied or is applied in a cost-effective and cost-ineffective ways (adapted from Frangopol and Kim 2019).

The relation between the structural performance threshold P_{th} and the total life-cycle cost C_{life} of an existing bridge is illustrated in Figure 1.4. Based on this figure, the relations between P_{th} and C_{life}, when the SHM is not applied or is applied in cost-effective and cost-ineffective ways, are compared in Figure 3.3 (Frangopol and Kim 2019). The minimum total life-cycle cost C_{life}^* is achieved from the optimum threshold of structural performance P_{th}^*. Although the SHM may lead to reduction in the costs for maintenance and expected failure cost, an ineffective SHM can produce a negative benefit (i.e., $C_{ben,m} = C_{life}^* - C_{life,m1}^* < 0$). This is because the SHM requires additional costs for system installation, operation, management, and data acquisition and analysis. The effective SHM to obtain the positive benefit (i.e., $C_{ben,m} = C_{life}^* - C_{life,m1}^* > 0$) should be based on the successful integration of (a) optimum SHM system installation, operation and management, (b) accurate and reliable damage detection and identification, (c) efficient use of SHM data to improve the effectiveness and accuracy of life-cycle performance and service life prediction under uncertainty (Frangopol 2011; Frangopol and Soliman 2016; Frangopol and Kim 2019).

3.3 BRIDGE PERFORMANCE ASSESSMENT AND PREDICTION BASED ON INSPECTION AND MONITORING DATA

The data from inspection and monitoring can be used to predict the damage propagation and structural performance under uncertainty. The related probabilistic approaches have been developed since the last two decades (Brownjohn 2007; Frangopol et al. 2008a, 2008b; Strauss et al. 2008; Frangopol and Kim 2014a, 2014c, 2019; Okasha and Frangopol 2012; Okasha et al. 2010, 2011, 2012; Yan and Frangopol 2019). Based on the general formulation of state function of Eq. (1.3), the state function $g(t)$ to compute the time-dependent reliability can be expressed as

$$g(t) = R(t) - S(t) \tag{3.6}$$

where $R(t)$ and $S(t)$ represent the time-dependent random variables associated with resistance and load effect, respectively.

Depending on the type of monitoring data, loading conditions and resistance deterioration mechanisms to be considered, the state function of Eq. (3.6) can be modified (Kim and Frangopol 2020). The state function based on the stresses associated with the monitored strains induced by live load is defined as (Liu et al. 2009a, 2009b)

$$g(t) = \sigma_0(t) - (1 + e) \times \sigma(t) \tag{3.7}$$

where $\sigma_0(t)$ is the predefined stress limit at time t, which can be determined using the data from the controlled loading test and/or finite element analysis for bridges, e is the measurement error in the monitored data and/or finite element analysis, and $\sigma(t)$ is the stress induced by live load at time t. In general, the measurement error e is assumed to be a normally distributed random variable with the mean of zero and standard deviation (SD) of σ_e. Figure 3.4 shows the effect of measurement error e on probability of failure p_f. Assuming that the predefined stress limit σ_0 is normally distributed with the mean and SD of 25 MPa and 1 MPa, respectively (denoted as N(25 MPa; 1 MPa), and the live load induced stress σ follows a normal distribution defined as N(15 MPa; 2 MPa), the associated PDFs of σ_0 and σ are provided in Figure 3.4(a). When the SDs of measurement error σ_e are equal to 0.2, 0.4 and 0.6, the PDFs of values of the state function defined in Eq. (3.7) are shown in Figure 3.4(b). Considering that the area under the PDF upper bounded by the vertical axis corresponds to the probability of failure p_f, it can be seen that an increase in σ_e leads to an increase in p_f. By applying higher quality of SHM systems, σ_e can be reduced.

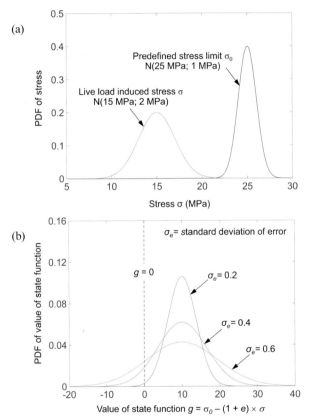

Figure 3.4 Effect of measurement error on probability of failure: (a) PDFs of predefined stress limit σ_0 and stress induced by live load σ; (b) PDFs of value of state function (see Eq. (3.7)) with different standard deviation of error σ_e.

The time-dependent stress induced by lived load (e.g., heavy vehicles) $\sigma(t)$ in Eq. (3.7) can be predicted as (Liu et al. 2009a, 2009b)

$$\sigma(t) = \sigma_{mon} \times \eta(t) \tag{3.8}$$

where σ_{mon} = stresses associated with the monitored strains; and $\eta(t)$ = time-dependent function which is defined as the ratio of the expected largest stress during the future time period t to the largest stress associated with the monitored strain. The time-dependent function $\eta(t)$ is defined as

$$\eta(t) = max\left[\frac{\sigma_{max,N_t}}{max\left\{\sigma_{mon,1}, \sigma_{mon,2}, ..., \sigma_{mon,n_t}\right\}}; 1.0\right] \qquad (3.9)$$

where $\sigma_{mon,i}$ = stress associated with the monitored strain induced by the ith heavy vehicle; n_t = total number of heavy vehicles during monitoring period; and σ_{max,N_t} = largest predicted stress among the stresses induced by N_t heavy vehicles during the future time period t. If the stress σ_{mon} is represented by the Type I asymptotic form, the largest predicted stress σ_{max,N_t} is computed as

$$\sigma_{max,N_t} = \alpha_{max} - \frac{\ln\left[-\ln\left(1-1/N_t\right)\right]}{\gamma_{max}} \qquad (3.10)$$

where α_{max} and γ_{max} are the location and scale parameters, respectively.

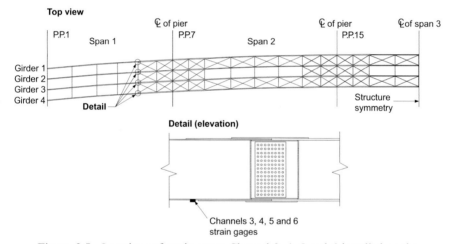

Figure 3.5   Locations of strain gages Channel 3, 4, 5 and 6 installed on the I-39 Northbound Wisconsin Bridge (adapted from Mahmoud et al. 2005).

For example, the monitored data of an existing bridge, Bridge I-39 Northbound Bridge located in Wisconsin, are considered. The monitored data were collected from the strain gages (i.e., Channels 3, 4, 5 and 6) installed on the bottom face of the bottom flange of each girder as shown in Figure 3.5. Figure 3.6 shows the histograms and the best-fit PDF distributions of the stress associated with the strain monitored from Channel 3 under the uncontrolled loadings for the right and left lanes. In order to compute the reliability of each girder considering the right, left, and side-by-side lane loadings, the state function of Eq. (3.7) is modified as

$$g(t) = \sigma_0 - (1+e) \times \eta(t) \times \left(\frac{N_r}{N_t} \cdot \sigma_{mon,r} + \frac{N_l}{N_t} \cdot \sigma_{mon,l} + \frac{N_{ss}}{N_t} \cdot \sigma_{mon,ss}\right) \qquad (3.11)$$

where $\sigma_{mon,r}$, $\sigma_{mon,l}$ and $\sigma_{mon,ss}$ are the stresses associated with the monitored strains induced by the right, left, and side-by-side lane loadings, respectively. N_r, N_l and N_{ss} are the total number of heavy vehicles crossing the bridge on the right, left, and side-by-side lanes, respectively, during a future time period t. The sum of N_r, N_l and N_{ss} is equal to N_t. Based on the state function defined in Eq. (3.11), the reliability of each girder can be obtained. It should be noted that the reliability based on Eq. (3.11) indicates the probability that the stress associated with the monitored strain induced by heavy vehicles is less than the predefined stress limit at time t.

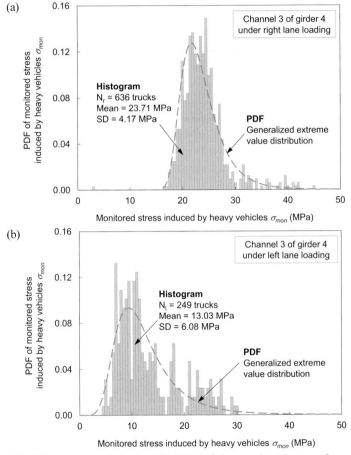

Figure 3.6 Histograms and the best-fit PDFs of the monitored stress from Channel 3 under the uncontrolled loadings: (a) right lane; (b) left lane (adapted from Liu et al. 2009b).

For fatigue reliability assessment and prediction, the state function considering the monitoring data can be expressed as (Kwon and Frangopol 2010, 2011; Guo et al. 2012; Kwon et al. 2013)

$$g(t) = \Delta - (1 + e) \cdot \frac{N_{total}}{A} \cdot S_{re}^m \qquad (3.12)$$

where Δ is the Miner's critical damage index, N_{total} is the cumulative number of cycles at time t, A is the fatigue detail coefficient, S_{re} is the effective stress range, and m is a constant. Using the monitoring data, the effective stress range S_{re} and the cumulative number of cycles N_{total} can be estimated. Assuming that Δ, e, A and S_{re} in Eq. (3.12) are lognormal random variables, the fatigue reliability index β can be computed as (Kwon and Frangopol 2010)

$$\beta = \frac{\lambda_{\Delta} + \lambda_A - (\lambda_e + m \cdot \lambda_{S_{re}} + \ln N_{total})}{\sqrt{v_{\Delta}^2 + v_A^2 + v_e^2 + (m \cdot v_{S_{re}})^2}} \qquad (3.13)$$

where λ_{Δ}, λ_e, λ_A, and λ_{Sre} are the means of $\ln(\Delta)$, $\ln(e)$, $\ln(A)$, and $\ln(S_{re})$, respectively. v_{Δ}, v_e, v_A, and v_{Sre} denote the SDs of $\ln(\Delta)$, $\ln(e)$, $\ln(A)$, and $\ln(S_{re})$, respectively. As a result, the time-dependent fatigue reliability index β is obtained as shown in Figure 3.7, where the monitoring data collected at Channels 9 of the Neville Island Bridge over the Ohio River, Pennsylvania, are used. The monitoring of this bridge was performed during 29 days from March to April 2004. The detailed description of this monitoring can be found in Connor et al. (2005). More information on the approach for the bridge fatigue reliability is available in Kwon and Frangopol (2010).

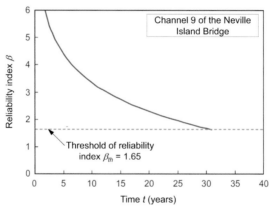

Figure 3.7 Time-dependent fatigue reliability index based on monitoring data collected at Channel 9 of the Neville Island Bridge (adapted from Kwon and Frangopol 2010).

As described in Section 2.2.4, the information from inspection and SHM can be integrated with the existing information on bridge performance by using the Bayes' theorem-based updating techniques. The random variables associated with the load effect and resistance of the state function can be updated, and finally, the bridge reliability prediction is updated. Figure 3.8 shows the existing and updated PDFs of the parameters (i.e., material parameter C, and stress range S_{re}) involved in fatigue crack propagation model of Eq. (2.14). The updated PDFs in Figure 3.8 are obtained using the cascaded Metropolis–Hastings sampling and slice sampling methods with detected fatigue cracks. Based on these PDFs, the time to reach the critical crack size t_{crt} can be re-estimated. Figure 3.9 compares the time to reach the critical crack size t_{crt} before and after updating. The details associated with Figures 3.8 and 3.9 are provided in Kim et al. (2019).

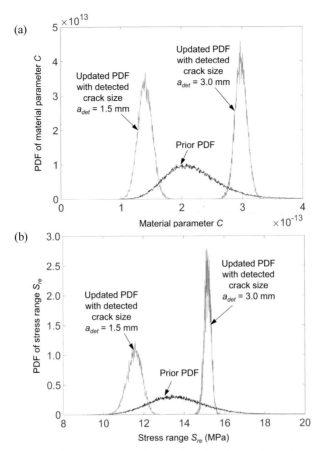

Figure 3.8 PDFs of parameters updated with the detected crack size: (a) material parameter C; (b) stress range S_{re} (adapted from Kim et al. 2019).

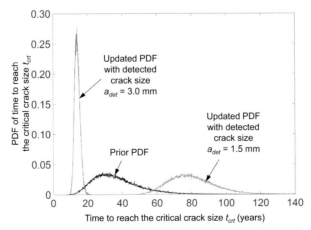

Figure 3.9 PDFs of time to reach the critical crack size before and after updating (adapted from Kim et al. 2019).

Table 3.1 Summary of investigations for bridge performance prediction and updating using inspection and monitoring data

Approaches	References
Use of inspection and monitoring data to assess and predict structural performance • Formulation of state function based on inspection and monitoring data • Finite element analysis with monitoring data • Structural performance assessment and prediction based on inspection and monitoring data • Effect of uncertainties associated with inspection and monitoring data on structural performance prediction	• Catbas et al. (2008) • Frangopol et al. (2008a, 2008b) • Strauss et al. (2008) • Marsh and Frangopol (2008) • Liu et al. (2009a, 2009b, 2010a, 2010b) • Kwon and Frangopol (2010, 2011) • Messervey et al. (2011) • Guo et al. (2012) • Okasha et al. (2010, 2011, 2012) • Okasha and Frangopol (2012) • Orcesi and Frangopol (2010, 2011a, 2013) • Karandikar et al. (2012) • Soliman et al. (2013b)
Use of inspection and monitoring data to update existing structural performance prediction • Updating parameters for PDFs representing resistance and load effect • Updating parameters for damage propagation model • Updating structural performance and service life prediction model	• Beck and Katafygiotis (1998) • Enright and Frangopol (1999a) • Estes and Frangopol (2003) • Strauss et al. (2008) • Straub, D. (2009) • Suo and Stewart (2009) • Karandikar et al. (2012) • Okasha et al. (2010, 2012) • Okasha and Frangopol (2012) • Maljaars and Vrouwenvelder (2014) • Thanapol et al. (2016) • Kim et al. (2019)

Table 3.1 provides a summary of the several approaches developed for bridge performance prediction and updating using inspection and monitoring data. In addition to the investigations presented previously, reliability of an existing long-span cantilever truss bridges based on long-term monitoring data was assessed by Catbas et al. (2008), where the finite element analysis is used with monitoring data to compute the reliability indices of components and system according to the framework indicated in Figure 3.10. Liu et al. (2010a, 2010b) investigated the fatigue reliability assessment and prediction of a retrofitted steel bridge (Birmingham Bridge located in Pittsburg, Pennsylvania). The monitoring data were collected from strain sensors installed near the retrofitted locations. The fatigue reliability of the retrofitted locations is estimated based on the finite element analysis with monitoring data. Guo et al. (2012) conducted the probabilistic finite element analysis with the weigh-in-motion data for fatigue reliability assessment of the Throgs Neck Bridge in New York City. Bayesian statistical framework to update the structural models using inspection and monitoring data was developed by Beck and Katafygiotis (1998). Enright and Frangopol (1999a) investigated the corrosion propagation and its updating based on inspection data for Colorado Highway Bridge L-18-BG. The probabilities of failure prediction before and after updating

were compared. Okasha et al. (2012) addressed the automated finite element updating using monitoring data. Based on their investigation, a general framework to integrate SHM data into bridge performance assessment and prediction was developed by Okasha and Frangopol (2012). Finally, a probabilistic procedure to estimate life-cycle seismic reliability of corroded existing RC structures based on inspection data was investigated by Thanapol et al. (2016).

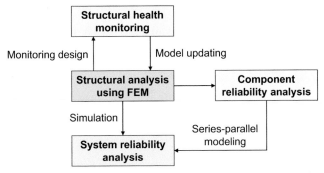

Figure 3.10 Simplified framework for reliability analysis using SHM data (adapted from Catbas et al. 2008).

3.4 AVAILABILITY OF MONITORING DATA FOR PERFORMANCE PREDICTION

The availability of a structural component (or system) is the probability that the component (or system) is functioning at time t. Similarly, the availability of monitoring data for bridge performance prediction can be defined as the probability that the prediction model for bridge performance is usable at a future time t_f (Kim and Frangopol 2011a). The criterion for usability of monitoring data for prediction is based on the difference between the values from the bridge performance prediction model and monitoring data. If the largest difference in the future time t_f exceeds the largest difference during the monitoring time period t_{md}, the monitoring data is unusable to predict the bridge performance. The associated exceedance probability p_{exd} is estimated based on Eq. (1.33) as

$$p_{exd}(t_f) = 1 - \exp\left(-\frac{t_f}{t_{md}}\right) \qquad (3.14)$$

When the monitoring data is usable over the prediction duration of t_f days, the availability of the monitoring data during t_f is 1.0. The average availability A' of monitoring data during the future time t_f is defined as (Ang and Tang 1984)

$$A' = \frac{t_l}{t_f} \cdot p_{exd}(t_f) + (1 - p_{exd}(t_f)) \qquad (3.15)$$

where t_l is the time to lose the usability of monitoring data for prediction. The expected average availability $E(A')$ for the monitoring duration t_{md} is expressed as

$$E\left(A'\right) = \frac{t_{md}}{t_f} \cdot \left(1 - \exp\left(-\frac{t_f}{t_{md}}\right)\right) \tag{3.16}$$

Figure 3.11 shows the exceedance probability p_{exd} and expected average availability of monitoring $E(A')$ defined in Eqs. (3.14) and (3.16), respectively, when the monitoring duration t_{md} is equal to 100 days. This figure shows that an increase in the prediction duration t_f leads to an increase in p_{exd} and decrease in $E(A')$. Based on maximizing the expected average availability $E(A')$, monitoring planning can be optimized (Kim and Frangopol 2011a).

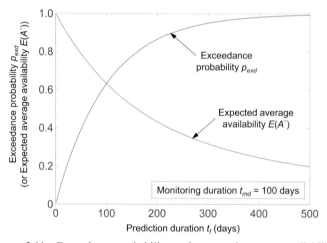

Figure 3.11   Exceedance probability and expected average availability.

3.5   DAMAGE DETECTION UNDER UNCERTAINTY

The uncertainty associated with damage detection exists, since the damage occurrence and propagation are under uncertainty, and inspection and monitoring cannot be perfect (Kim and Frangopol 2011a, 2011b, 2011c, 2011d). For this reason, the damage detection time during the service life of a deteriorating structure should be estimated using probabilistic concepts and methods. Based on the probabilistic estimation of damage detection time and delay, the optimum inspection, monitoring and maintenance planning can be achieved (Frangopol and Kim 2019).

3.5.1   Probability of Damage Detection for Multiple Inspections

The probability of damage detection for a single inspection application is described in Eqs. (3.1) to (3.4). Considering the relation among the inspection application times, service life and probability of damage detection, the probability of damage detection PoD_N for N_{ins} inspections is formulated as (Soliman et al. 2013a)

$$PoD_N = \sum_{j=1}^{N_{ins}} \left[\prod_{i=1}^{j} \left\{ P\left(t_{ins,i} \le t_{life}\right) \cdot \left(1 - P_{det,i-1}\right) \right\} \cdot P_{det,j} \right] \quad (3.17)$$

where $P_{det,j}$ = probability of damage detection of the jth inspection applied at the time $t_{ins,j}$; and $P(t_{det,i} \le t_{life})$ = probability that the ith inspection is applied before the end of service life t_{life}. The probability $P_{det,i-1}$ for $i = 1$ is considered as zero (i.e., $P_{det,0} = 0$). Figure 3.12 shows the event tree to formulate the probability of damage detection PoD_2 of Eq. (3.17) for two inspections at time $t_{ins,1}$ and $t_{ins,2}$. In Figure 3.12, Branches 2 and 3 correspond to the damage detection at the first and second inspections, respectively. The occurrence probability associated with Branch 2 is $P(t_{ins,1} \le t_{life}) \times P_{det,1}$. Branch 3 has the occurrence probability $P(t_{ins,1} \le t_{life}) \times P(t_{ins,2} \le t_{life}) \times (1 - P_{det,1}) \times P_{det,2}$. Therefore, the probability of damage detection PoD_2 for two inspections becomes

$$PoD_2 = P(t_{ins,1} \le t_{life}) \cdot P_{det,1} + P(t_{ins,1} \le t_{life}) \cdot P(t_{ins,2} \le t_{life}) (1 - P_{det,1}) \cdot P_{det,2} \quad (3.18)$$

In a similar way, the formulation of PoD_N for N_{ins} inspections can be obtained as indicated in Eq. (3.17). Increasing the number of inspections and/or probability of damage detection of an inspection method can lead to an increase in the probability of damage detection PoD_N of Eq. (3.17).

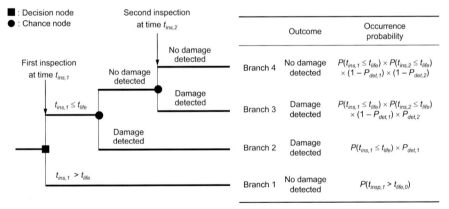

Figure 3.12 Event tree to formulate the probability of damage detection associated with two inspections.

3.5.2 Damage Detection Time and Delay for Inspection and Monitoring

The relation among the damage detection time t_{det}, damage detection delay t_{del}, and damage occurrence time t_{oc} is shown in Figure 3.13, which is expressed as

$$t_{det} = t_{del} + t_{oc} \qquad \text{for} \qquad t_{det} \ge t_{oc} \quad (3.19)$$

The damage occurrence time t_{oc} is associated with the time when the degree of damage reaches a predefined threshold. The damage detection delay is formulated by using the event tree model (Kim and Frangopol 2011c, 2011d). When N_{ins} inspections are performed at times $t_{ins,1}$, $t_{ins,2}$, ... and $t_{ins,N}$, the formulation of the damage detection time $t_{det,i}$ for damage occurrence in the time interval between $t_{ins,i-1}$ and $t_{ins,i}$ is

$$t_{det,i} = \sum_{k=i}^{N_{ins}+1} \left\{ \left(\prod_{j=1}^{k} \left(1 - P_{det,j-1}\right) \right) \cdot P_{det,k} \cdot t_{ins,k} \right\} \quad \text{for } t_{ins,i-1} \leq t_{oc} < t_{ins,i} \quad (3.20)$$

where $P_{det,0}$ for $j = 1$ and $P_{det,N_{ins}+1}$ for $k = N_{ins}+1$ are zero and one, respectively. Based on Eq. (3.19), the damage detection delay $t_{del,i}$ for $t_{ins,i-1} \leq t_{oc} < t_{insp,i}$ is

$$t_{del,i} = \sum_{k=i}^{N_{ins}+1} \left\{ \left(\prod_{j=1}^{k} \left(1 - P_{det,j-1}\right) \right) \cdot P_{det,k} \cdot t_{ins,k} \right\} - t_{oc} \quad \text{for } t_{ins,i-1} \leq t_{oc} < t_{ins,i} \quad (3.21)$$

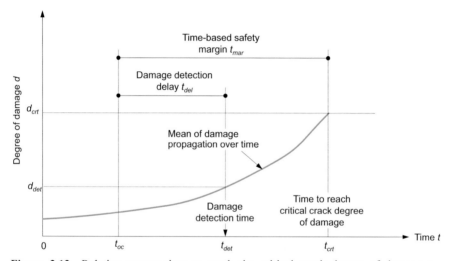

Figure 3.13 Relation among time to reach the critical crack degree of damage t_{crt}, damage detection time t_{det}, time-based safety margin t_{mar} and damage detection delay t_{del}.

The uncertainty associated with the damage occurrence time t_{oc}, which is represented by the PDF $f_T(t_{oc})$, is integrated into the formulation of the expected damage detection time $E(t_{det})$ and expected damage detection delay $E(t_{del})$ as (Kim and Frangopol 2011c and 2011d)

$$E\left(t_{det}\right) = \sum_{i=1}^{N_{ins}+1} \left\{ \int_{t_{ins,i-1}}^{t_{ins,i}} \left[t_{det,i} \cdot f_T\left(t_{oc}\right) \right] dt_{oc} \right\} \quad (3.22a)$$

$$E\left(t_{del}\right) = \sum_{i=1}^{N_{ins}+1} \left\{ \int_{t_{ins,i-1}}^{t_{ins,i}} \left[t_{del,i} \cdot f_T\left(t_{oc}\right) \right] dt_{oc} \right\} \quad (3.22b)$$

$t_{ins,0}$ for $i = 1$ and $t_{ins,N_{ins}+1}$ for $i = N_{ins}+1$ are the lower bound t_{lb} and upper bound t_{ub} of damage occurrence time, respectively. These bounds are defined as

$$t_{lb} = F_T^{-1}[\Phi(-k_b)] \tag{3.23a}$$

$$t_{ub} = F_T^{-1}[\Phi(k_b)] \tag{3.23b}$$

where $F_T^{-1}(\cdot)$ = the inverse CDF of the damage occurrence time t_{oc}; and k_b = parameter for the time interval between t_{lb} and t_{ub}.

If the SHM systems installed in appropriate locations monitor continuously the changes in structural characteristics and behaviors of a deteriorating bridge during a long-term period, and the monitored data are interpreted accurately, the structural damage can be detected perfectly (Frangopol and Kim 2019). Assuming that there is no damage detection delay during the monitoring, the formulations of the expected damage detection time $E(t_{det})$ and expected damage detection delay $E(t_{del})$ defined in Eq. (3.22) are modified as (Kim and Frangopol 2011c, 2012)

$$E(t_{det}) = \sum_{i=1}^{N_{mon}+1} \left\{ \int_{t_{ms,i-1}+t_{md}}^{t_{ms,i}} \left[t_{ms,i} \cdot f_T(t_{oc}) \right] dt_{oc} + \int_{t_{ms,i}}^{t_{ms,i}+t_{md}} \left[t_{oc} \cdot f_T(t_{oc}) \right] dt_{oc} \right\} \tag{3.24a}$$

$$E(t_{del}) = \sum_{i=1}^{N_{mon}+1} \left\{ \int_{t_{ms,i-1}+t_{md}}^{t_{ms,i}} \left[(t_{ms,i} - t_{oc}) \cdot f_T(t_{oc}) \right] dt_{oc} \right\} \tag{3.24b}$$

where N_{mon} = number of monitorings; $t_{ms,i}$ = ith monitoring starting time; and t_{md} = monitoring duration. $t_{ms,i-1} + t_{md}$ for $i = 1$ and $t_{ms,i}$ for $N_{mon} + 1$ indicate t_{lb} and t_{ub} defined in Eq. (3.23), respectively.

Furthermore, the expected damage detection delay $E(t_{del})$ and expected damage detection time $E(t_{det})$ for combined inspection and monitoring are formulated based on Eqs. (3.22) and (3.24). For example, if one monitoring is applied after one inspection, the expected damage detection time $E(t_{det})$ and expected damage detection delay $E(t_{del})$ can be formulated as (Kim and Frangopol 2012)

$$E(t_{det}) = \int_{t_{lb}}^{t_{ins,1}} \left[P_{det,1} \cdot t_{ins,1} + (1 - P_{det,1}) \cdot t_{ms,1} \right] \cdot f_T(t_{oc}) dt_{oc}$$
$$+ \int_{t_{ins,1}}^{t_{ms,1}} t_{ms,1} \cdot f_T(t_{oc}) dt_{oc} + \int_{t_{ms,1}}^{t_{ms,1}+t_{md}} t_{oc} \cdot f_T(t_{oc}) dt_{oc} + \int_{t_{ms,1}+t_{md}}^{t_{ub}} t_{lb} \cdot f_T(t_{oc}) dt_{oc} \tag{3.25a}$$

$$E(t_{del}) = \int_{t_{lb}}^{t_{ins,1}} \left[P_{det,1} \cdot (t_{ins,1} - t_{oc}) + (1 - P_{det,1}) \cdot (t_{ms,1} - t_{oc}) \right] \cdot f_T(t_{oc}) dt_{oc}$$
$$+ \int_{t_{ins,1}}^{t_{ms,1}} (t_{ms,1} - t_{oc}) \cdot f_T(t_{oc}) dt_{oc} + \int_{t_{ms,1}+t_{md}}^{t_{ub}} (t_{ub} - t_{oc}) \cdot f_T(t_{oc}) dt_{oc} \tag{3.25b}$$

Additional descriptions on the formulation of $E(t_{del})$ and $E(t_{det})$ associated with more combinations of inspection and monitoring are provided in Kim and Frangopol (2012).

3.5.3   Damage Detection-Based Probability of Failure

If the damage detection leads to immediate repair, a failure criterion based on damage detection is expressed as (Kim and Frangopol 2011b)

$$t_{crt} - t_{det} = t_{mar} - t_{del} < 0 \qquad (3.26)$$

where t_{crt} = time to reach the critical degree of damage d_{crt}; and t_{mar} = time-based safety margin. Figure 3.13 shows the relation among t_{crt}, t_{det}, t_{mar} and t_{del}. Due to the uncertainties associated with damage occurrence and propagation, and damage detection, t_{crt}, t_{det}, t_{mar} and t_{del} can be treated as random variables. Therefore, the probability of failure p_f based on damage detection is defined as

$$p_f = P(t_{crt} - t_{det} < 0) = P(t_{mar} - t_{del} < 0) \qquad (3.27)$$

Figure 3.14 shows the PDFs of t_{crt}, t_{det}, t_{mar} and t_{del}. Area 1 in Figure 3.14(a) corresponds to the probability of failure $p_f = P(t_{crt} - t_{det} < 0)$. The probability of failure $p_f = P(t_{mar} - t_{del} < 0)$ is associated with Area 2 in Figure 3.14(b). According to Eq. (3.27), these two areas are equal (i.e., Area 1 = Area 2).

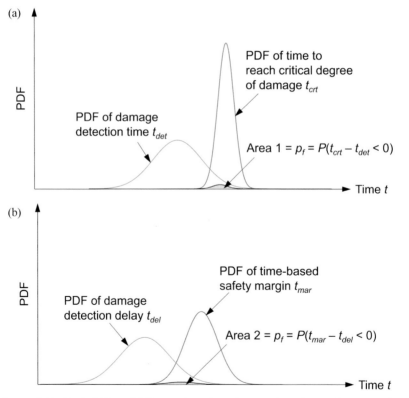

Figure 3.14   Probability of failure p_f based on damage detection under uncertainty: (a) $t_{crt} - t_{det}$; (b) $t_{mar} - t_{del}$.

The effects of probability of damage detection $P_{det,i}$ of an inspection method and number of inspections N_{ins} on expected damage detection delay $E(t_{del})$ and damage detection-based probability of failure p_f are illustrated in Figure 3.15(a) and 3.15(b), respectively. From these figures, it can be seen that increasing $P_{det,i}$ and/or N_{ins} results in reduction of $E(t_{del})$ and p_f. Figure 3.16 shows the effects of monitoring durations t_{md} and number of monitorings t_{md} on $E(t_{del})$ and p_f. Reduction of $E(t_{del})$ by increasing t_{md} and/or N_{mon} can be found in Figure 3.16(a). Figure 3.16(b) shows that an increase in t_{md} and/or N_{mon} can produce a decrease of p_f. More investigations related to Figures 3.15 and 3.16 are available in Kim and Frangopol (2011c).

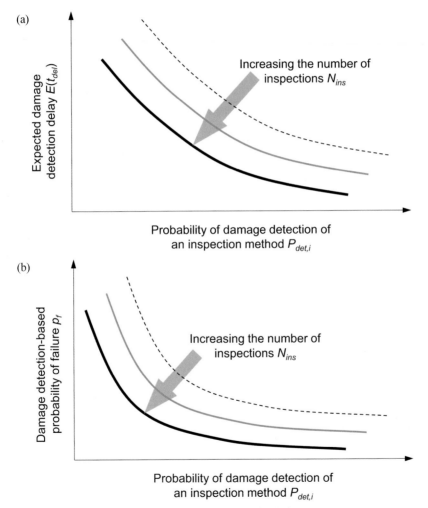

Figure 3.15 Effects of probability of damage detection of an inspection method and number of inspections on: (a) expected damage detection delay; (b) damage detection-based probability of failure.

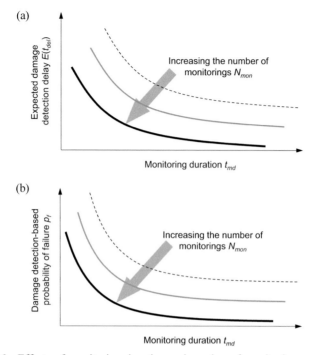

Figure 3.16 Effects of monitoring duration and number of monitorings on: (a) expected damage detection delay; (b) damage detection-based probability of failure.

3.6 CONCLUDING REMARKS

In this chapter, several representative inspection and SHM techniques for fatigue-sensitive steel bridges and RC bridges are reviewed. The probabilistic approaches to integrate the data from SHM into structural performance assessment and prediction are presented. The availability of monitoring data for structural performance prediction is described, where the effects of monitoring and prediction durations on the expected availability of monitoring data are provided. The uncertainties associated with the damage occurrence and propagation, and damage detection are addressed to formulate the expected damage detection time and delay, and probability of failure. The relations among the probability of damage detection of an inspection method, monitoring duration, and number of inspections and monitorings, expected damage detection delay and probability of failure are investigated. The expected damage detection delay and probability of failure can be reduced by increasing the probability of damage detection of an inspection method, monitoring duration, and number of inspections and/or monitorings. Based on the probabilistic concepts such as availability of monitoring data, damage detection time and delay, damage detection-based probability of failure, the inspection and monitoring planning can be optimized. The associated investigations will be presented in Chapter 6.

Chapter 4

Bridge Maintenance

NOTATIONS (continued)	
$t_{life,em}$	= extended service life with essential maintenance
$t_{life,em+pm}$	= extended service life with both essential and preventive maintenances
$t_{life,ex}$	= expected extended service life
$t_{life,pm}$	= extended service life with preventive maintenance
t_{em}	= application time for essential maintenance
t_{pd}	= duration of maintenance effect
t_{pm}	= application time for preventive maintenance
$t_{pm,d}$	= time interval between preventive maintenances
β	= reliability index
β_C	= reliability index of combined series-parallel system
β_P	= reliability index of parallel system
β_S	= reliability index of series system
β_{in}	= initial reliability index
β_{th}	= threshold of reliability index
$\Delta\beta$	= reliability index improvement after maintenance
Δr_β	= reduction of deterioration rate after maintenance
ρ_{ij}	= correlation coefficient between the service life extensions due to the ith and jth maintenances
$\rho_{(k)}$	= correlation matrix of the service life extensions by applying maintenances to the kth bridge component
$\rho_{(sys)}$	= correlation matrix of the service life extensions by applying maintenances to the bridge system

ABSTRACT

Chapter 4 presents the bridge maintenance types and their effects on performance, service life, service life extension and cost. The characteristics and representative types of preventive maintenance (PM), essential maintenance (EM) and replacement (RP) are presented. The relations among the bridge performance, service life, and maintenance at both component-level and system-level are investigated. The maintenance modelings based on the reliability profiles, lifetime functions, damage propagation and event trees are described. The effect of correlation among the service life extensions due to various maintenances on the extended service life of components and systems is investigated. The general process for bridge maintenance consisting of maintenance planning, maintenance application, reporting and evaluating after maintenance, and bridge management systems are presented.

4.1 INTRODUCTION

Maintenance of deteriorating bridges is essential to ensure their serviceability and safety under uncertainty. The purposes of bridge maintenance include prevention, delay and reduction of the bridge performance deterioration, improvement of the bridge condition and safety, and service life extension (FHWA 2015, 2018). Since the bridge maintenance requires cost and the financial resources are limited, the cost-effective bridge maintenance management should be investigated (AASHTO 2007a). Over the last several decades, the techniques for bridge maintenance and approaches related to maintenance management have been developed extensively due to a substantial increase in the number of structurally deficient bridges (NCHRP 2003, 2009). The bridge maintenance management has to be based on a proper understanding of the maintenance effects on bridge performance, service life and life-cycle cost (Sánchez-Silva et al. 2016). Finally, the well-balanced bridge maintenance management can be achieved through the optimization process (Frangopol 2011; Frangopol et al. 2017).

In this chapter, the effects of bridge maintenance on performance, service life, service life extension and cost are addressed. The characteristics and representative types of preventive maintenance (PM), essential maintenance (EM) and replacement (RP) are presented. The relations among the bridge performance, service life, and maintenance at both component-level and system-level are investigated. The approaches for maintenance modeling are presented. Using these approaches the effects of various maintenance types on the structural performance, extended service life and total life-cycle cost of deteriorating bridges can be formulated in a systematic way. The maintenance modelings based on the reliability profiles, lifetime functions, damage propagation and event trees are provided. The changes in reliability by applying maintenances are captured in the reliability profile. The survival functions with PM, EM and RP are formulated. The effect of maintenance on structural performance and cost is modeled based on damage propagation under uncertainty. The event tree for maintenance modeling considers the uncertainties associated with predictions of service life, damage detection, maintenance intervention and maintenance cost. The correlation of service life extensions due to various maintenance actions is investigated. Furthermore, the processes and systems for bridge management are presented.

4.2   BRIDGE MAINTENANCE AND REPLACEMENT

In general, bridge asset management indicates the actions and strategies related to the prevention of unexpected structural failure, service life extension, deterioration delay and reduction, and improvement of structural performance and condition (Frangopol et al. 2001; AASHTO 2007a; Alampalli 2014; FHWA 2018). As shown in Figure 4.1, bridge asset management is categorized into PM, EM (or rehabilitation), and RP (FHWA 2018). Figure 4.2 shows the time-dependent bridge conditions and appropriate bridge maintenance types. The four conditions (good, fair, poor and severe) in Figure 4.2 are related to the National Bridge Inventory (NBI) condition rating values from 0 to 9. The good condition is associated with the NBI condition rating values 7, 8 and 9. The NBI condition rating values 5 and 6 are equivalent to the fair condition. The poor condition is represented by the NBI condition rating value 4. The severe condition can be expressed with the NBI rating values from 0 to 3 (FHWA 2018). For good and fair bridge conditions, the PM can be applied at a scheduled time and/or predefined condition to reduce bridge performance deterioration. The EM is performed when the bridge performance reaches a fair or poor bridge condition. The bridge replacement applied at poor or severe bridge condition refers to the complete replacement of a structurally deficient bridge with a new bridge, which leads to an initial bridge performance (Alampalli 2014; FHWA 2018).

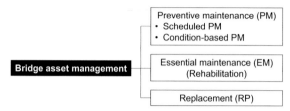

Figure 4.1   Bridge asset management categories (adapted from FHWA 2018).

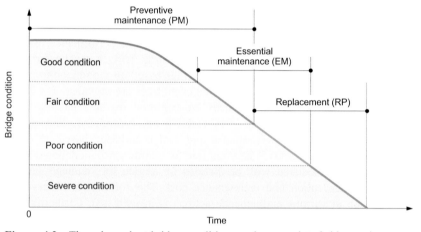

Figure 4.2   Time-dependent bridge conditions and appropriate bridge maintenance types (adapted from FHWA 2018).

4.2.1   Preventive Maintenance

The PM is applied at a predefined time interval to keep a bridge in good or fair condition, to delay structural performance deterioration, and to prevent the enormous future costs caused by bridge rehabilitation or replacements (FHWA 2015). For example, the damages in steel and concrete bridges such as corrosion and corrosion-induced cracking can be avoided by applying PM to protect a bridge from chloride-induced ingression and to provide adequate drainage. As a result, bridge deterioration can be delayed and reduced. The associated maintenance cost is much less than the cost for replacement of concrete deck (Fisher et al. 1998; Sprinkel 2001). The PM applied at a scheduled time interval includes (a) cleaning decks, super- and sub-structures, (b) cleaning expansion joints, (c) cleaning bridge drainage systems, and (d) sealing concrete decks, super- and sub-structures. When the damage is detected, or the inspected condition is below the threshold, the condition-based PM including (a) resealing, repairing and replacing expansion joints, (b) sealing, patching and repairing concrete decks, super-and sub-structures, (c) painting steel members, and (d) fatigue crack mitigation can be applied. Additional types of PM and associated descriptions are provided in AASHTO (2007a), NYSDOT (2008) and FHWA (2018). The PM planning can be optimized considering the effective life and cost of the PM, and service life extension after the PM. Table 4.1 provides the minimum and average service life extension of the bridge components through the PM (Yanev 2003; Alampalli 2014).

Table 4.1 Minimum and average service life of bridge components to be maintained using PM (adapted from Yanev 2003 and Alampalli 2014)

Bridge components	Minimum service life (years)	Average service life (years)
Abutments	35	100
Backwalls	35	80
Bearings	20	60
Bridge seats	20	80
Curbs	15	70
Piers	30	70
Joints	10	60
Primary member	30–35	80
Secondary member	35	80
Sidewalks	15	70
Wearing surfaces	10–35	50–60
Wing walls	50	80

4.2.2 Essential Maintenance

The EM performed at the time when the bridge performance reaches a fair or poor condition can result in restoring bridge integrity, increasing the bridge performance, and extending the service life. The inspection of a bridge in a fair or poor condition is generally followed by the EM. The EM includes partial or complete replacement of deck, superstructure and piers (FHWA 2018). If the bridge component is replaced by the EM, the performance of the bridge component can be restored to its original state (Okasha and Frangopol 2010b). The cost-effective EM requires a careful remaining service life estimation of bridge components and/or entire bridge (Alampalli 2014). For example, a bridge associated with the predicted remaining service life of 20 years may not need the deck replacement, which can last more than 30 years.

4.2.3 Replacement

Maintenance is the bridge component-level work to improve the condition and performance of the component without complete restoration of the bridge system to the initial condition and performance (Alampalli 2014). According to FHWA (2018), RP is defined as the complete replacement of an existing bridge with a new one constructed in the same traffic corridor. When the bridge condition is poor or severe, and maintenance is not cost-effective, and cannot significantly improve the bridge condition, RP needs to be applied. RP cost is much higher than the cost of PM or EM. Since the replacement of a bridge should be based on the current design standards and requirements, the bridge after RP will have at least its original performance and service life.

4.2.4 Effects of Maintenance and Replacement on Performance, Service Life and Cost of Bridge Component and System

The maintenance and replacement of bridge components can improve the bridge performance, and extend the service life. However, these actions can be very costly. Under limited financial resources, the cost for maintenance and replacement of bridge components should be allocated properly (Frangopol 2011; Frangopol and Soliman 2016; Frangopol et al. 2017). Figure 4.3 shows the relation among the bridge performance, service life, and maintenance at a component-level. The PM applied at time t_{pm} leads to the service life extension from $t_{life,0}$ to $t_{life,pm}$, while the PM requires the maintenance cost $C_{ma,pm}$. When the bridge component performance after the PM reaches the performance threshold P_{th} at the time $t_{life,pm}$, the EM requiring the additional cost $C_{ma,em}$ can result in the service life extension from $t_{life,pm}$ to $t_{life,pm+em}$.

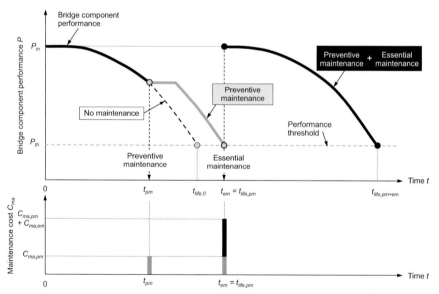

Figure 4.3 Time-dependent bridge performance and cumulative maintenance cost at a component-level.

Based on Figure 4.3, the relation among the bridge performance (i.e., reliability index), service life, and maintenance at a system-level is illustrated in Figure 4.4,

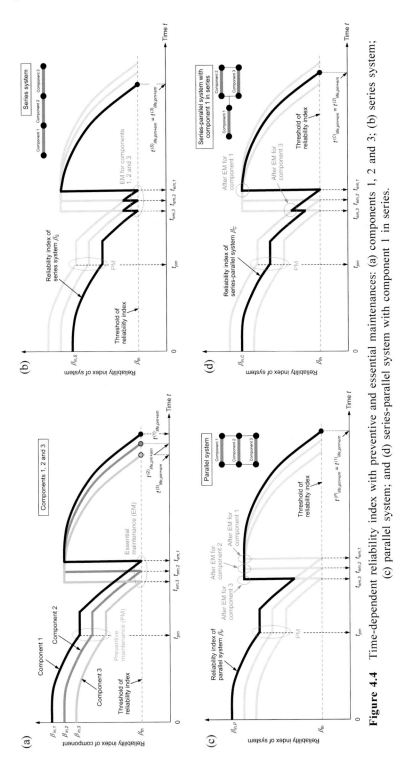

Figure 4.4 Time-dependent reliability index with preventive and essential maintenances: (a) components 1, 2 and 3; (b) series system; (c) parallel system; and (d) series-parallel system with component 1 in series.

where the series, parallel and series-parallel systems consisting of three components 1, 2 and 3 are considered. Figure 4.4(a) provides the time-dependent reliability indices of three components 1, 2 and 3, assuming:

- The initial reliability index of component 1 is larger than those of the components 2 and 3, and the initial reliability index of component 2 is larger than that of component 3 (i.e., $\beta_{in,1} > \beta_{in,2} > \beta_{in,3}$)
- The three components 1, 2 and 3 have the same deterioration initiation time and rate.
- The PM application time t_{pm} and its effect on the reliability index of components are the same for the three components.
- The EM is applied when the reliability index β reaches the threshold reliability index β_{th}, and its effect on the reliability index of components are the same for the three components.
- The maintenance costs of the PM and EM for the three components are the same ($C^{(1)}_{ma,pm} = C^{(2)}_{ma,pm} = C^{(3)}_{ma,pm} = C_{ma,pm}$, and $C^{(1)}_{ma,em} = C^{(2)}_{ma,em} = C^{(3)}_{ma,em} = C_{ma,em}$, where $C^{(i)}_{ma,pm}$ and $C^{(i)}_{ma,em}$ are the maintenance costs for the PM and EM, respectively).

As a result, the extended service lives of components 1, 2, and 3 become $t^{(1)}_{life,pm+em}$, $t^{(2)}_{life,pm+em}$, and $t^{(3)}_{life,pm+em}$, respectively, as shown in Figure 4.4(a).

If the safety margins of components 1, 2 and 3 in Figure 4.4(a) are perfectly positive correlated, the reliability index profiles of the series, parallel, and series-parallel systems consisting of these three components are shown in Figures 4.4(b), 4.4(c) and 4.4(d), respectively. Based on Eq. (1.14), the reliability indices of these three systems are

$$\beta_S = \min_{i=1}^{3}\{\beta_i\} \qquad \text{for a series system} \qquad (4.1a)$$

$$\beta_P = \max_{i=1}^{3}\{\beta_i\} \qquad \text{for a parallel system} \qquad (4.1b)$$

$$\beta_C = \min\left[\beta_1, \max\{\beta_2, \beta_3\}\right] \qquad \begin{array}{l}\text{for a series-parallel system with} \\ \text{component 1 in series}\end{array} \qquad (4.1c)$$

where β_S, β_P and β_C are the reliability indices of the series, parallel and combined series-parallel system, respectively. β_i is the reliability index of the ith component. The assumption of perfectly positive correlation will produce a maximum reliability of series systems and a minimum reliability of parallel systems. The series, parallel, and series-parallel systems in Figures 4.4(b), 4.4(c) and 4.4(d) require three-time PMs and EMs, and the total maintenance cost of $C_{ma} = 3 \cdot C_{ma,pm} + 3 \cdot C_{ma,em}$. From the comparison among the reliability profiles in Figures 4.4(b), 4.4(c) and 4.4(d), the following is found:

- Although the three systems in Figures 4.4(b), 4.4(c) and 4.4(d) require the same total maintenance cost $C_{ma} = 3C_{ma,pm} + 3C_{ma,em}$, the extended service lives of the three systems are not the same. The extended service life of the series system $t^{(S)}_{life,pm+em}$ is shortest, and the parallel system has the largest extended service life $t^{(P)}_{life,pm+em}$. The extended service life of the series-parallel system $t^{(C)}_{life,pm+em}$ is between $t^{(S)}_{life,pm+em}$ and $t^{(P)}_{life,pm+em}$.

- As shown in Figure 4.4(b) and Eq. (4.1a), the reliability index of the series system β_S is the smallest reliability index among the reliability indices β_1, β_2 and β_3, before and after the PM and EM are applied. For example, before the first EM is applied at time $t_{em,3}$, the reliability index of the series system β_S is represented by the reliability index of component 3, β_3, which is the smallest among the three-component reliability indices β_1, β_2 and β_3. Finally, the extended service life of the series system is equal to the extended service life of component 3 (i.e., $t^{(S)}_{life,pm+em} = t^{(3)}_{life,pm+em}$).

- For the parallel system in Figure 4.4(c), the reliability index of the parallel system β_P is the same as the largest reliability index among β_1, β_2 and β_3 (see Eq. (4.1b)). For example, the reliability index of component 1, β_1, which is the largest one among β_1, β_2 and β_3, represents the reliability index of the parallel system β_P before the EM for component 3 is applied at time $t_{em,3}$. After the EM for component 1 is applied at time $t_{em,1}$, β_P is represented by β_1. As a result, the extended service life of the parallel system becomes the extended service life of component 1 (i.e., $t^{(P)}_{life,pm+em} = t^{(1)}_{life,pm+em}$).

- For the series-parallel system with component 1 in series in Figure 4.4(d), the system reliability index β_C is equal to the reliability index of component 2, β_2, before $t_{em,3}$. The first EM is applied on component 3 at time $t_{em,3}$. After that, β_C is equal to the reliability index of unmaintained component 1, β_1. When the EM is performed for component 1, at time $t_{em,1}$, the system reliability β_C is represented by the reliability index of component 2, β_2. The final extended service life of the series-parallel system is the same as the extended service life of component 2 (i.e., $t^{(C)}_{life,pm+em} = t^{(2)}_{life,pm+em}$).

4.3 BRIDGE MAINTENANCE MODELING

To find out in a systematic way the effects of the maintenance of components on the structural performance (e.g., reliability, redundancy, vulnerability and robustness), extended service life and total life-cycle cost of a bridge, appropriate maintenance modeling is essential. With the maintenance model, the maintenance planning to determine the times and types of maintenance actions can be optimized. The maintenance modeling can be based on the performance profiles, lifetime functions, damage propagation and event trees (Frangopol et al. 1997b; Enright and Frangopol 1999b; van Noortwijk and Frangopol 2004; Liu and Frangopol 2005b; Neves and Frangopol 2005; Yang et al. 2006a, 2006b; Frangopol and Liu 2007b; Kim et al. 2013).

4.3.1 Maintenance Modeling Based on Performance Profiles

The deterioration profile using linear and nonlinear models can represent the time-dependent structural performance with and without maintenance interventions. Frangopol et al. (2001) introduced the multilinear reliability index profiles considering the effects of PM and EM on the improvement of the reliability index and service life extension, where the parameters to represent the reliability profile are treated as random variables as shown in Figure 1.2. For example, the bridge reliability index without maintenance can be expressed as (Frangopol et al. 2001)

$$\beta(t) = \beta_{in} \qquad \text{for} \qquad t < t_{ini} \qquad (4.2a)$$

$$\beta(t) = \beta_{in} - r_{\beta}(t - t_{ini}) \qquad \text{for} \qquad t \geq t_{ini} \qquad (4.2b)$$

where t_{ini} = deterioration initiation time; β_{in} = initial reliability index; and r_{β} = deterioration rate. Based on the linear reliability index profile shown in Figure 4.5, the reliability index at the time t' since the ith maintenance application is (Frangopol et al. 2001; Liu and Frangopol 2005b)

$$\beta(t') = \beta_i(0) + \Delta\beta \qquad \text{for} \qquad 0 \leq t' < t_d \qquad (4.3a)$$

$$\beta(t') = \beta_i(0) + \Delta\beta - max\{r_{\beta} - \Delta r_{\beta}; 0\} \cdot (t' - t_d) \qquad \text{for} \quad t_d \leq t' < t_{pd} \qquad (4.3b)$$

$$\beta(t') = \beta_i(0) + \Delta\beta - max\{r_{\beta} - \Delta r_{\beta}; 0\} \cdot (t_{pd} - t_d) - r_{\beta}(t' - t_{pd}) \qquad \text{for } t_{pd} \leq t' \qquad (4.3c)$$

where $\beta_i(0)$ is the reliability index before the ith maintenance application, $\Delta\beta$ is the reliability index improvement, Δr_{β} is the reduction of deterioration rate after maintenance, t_d is the time delay of deterioration, and t_{pd} is the duration of maintenance effect.

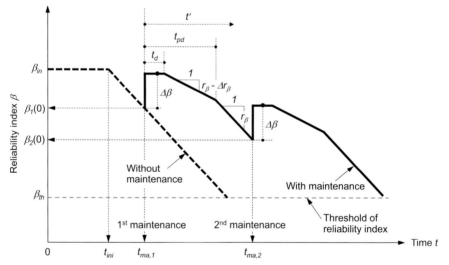

Figure 4.5   Linear reliability index profile with maintenance interventions (adapted from Frangopol et al. 2001 and Liu and Frangopol 2005b).

Considering multiple failure modes and the reliability improvements caused by the multiple maintenance actions, the reliability index associated with the jth failure mode at time t can be generalized as (Kong and Frangopol 2003b)

$$\beta^{(j)}(t) = \beta_0^{(j)}(t) + \sum_{i=1}^{N_{ma}} \Delta\beta_i^{(j)}(t_i) \qquad (4.4)$$

where $\beta_0^{(j)}(t)$ = reliability index without maintenance for the jth failure mode at time t; N_{ma} = number of maintenance actions; and $\Delta\beta_i^{(j)}(t_i)$ = reliability index improvement for the jth failure mode after the ith maintenance action is applied at time t_i. The six representative cases of the profile for $\Delta\beta(t)$ are provided in Figure 4.6(a), where t_s, t_i and t_e denote the starting, interim and ending times of the maintenance action, and $\Delta\beta_s$ and $\Delta\beta_e$ are the starting and ending improvement of reliability index, respectively. As an example, the effect of $\Delta\beta(t)$ associated with Case E in Figure 4.6(a) on the reliability index profile associated with no maintenance is illustrated in Figure 4.6(b). The system reliability profile can be obtained by probabilistic integration of the reliability profiles of bridge components under various failure modes.

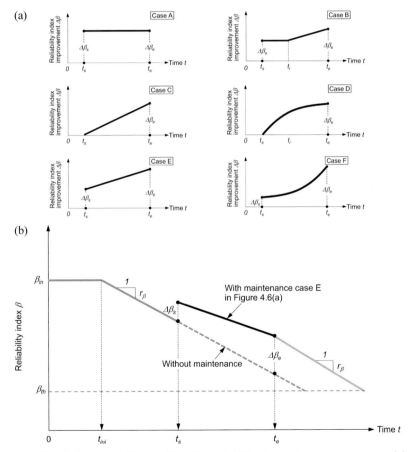

Figure 4.6 (a) Representative profiles for reliability index improvement caused by a maintenance intervention; and (b) reliability index profile integrating reliability index improvement profiles (adapted from Kong and Frangopol 2004b).

The expected total maintenance cost $E(C_{ma})$ is formulated as (Kong and Frangopol 2003b)

$$E\left(C_{ma}\right) = \sum_{i=1}^{N_{ma}} \left[\int_{t_{l,i}}^{t_{u,i}} \frac{C_{ma,i}}{\left(1+r_{dis}\right)^{t_i}} f_T\left(t_i\right) dt_i \right] \qquad (4.5)$$

where $C_{ma,i}$ is the ith maintenance cost, r_{dis} is the discount rate of money, $t_{l,i}$ and $t_{u,i}$ are the lower and upper bounds of the ith maintenance application time t_i, respectively, and $f_T(t)$ is the probability density function (PDF) of the maintenance application time. Based on the relation between maintenance cost and reliability improvement resulted from a maintenance intervention, the ith maintenance cost $C_{ma,i}$ in Eq. (4.5) can be expressed as (Kong and Frangopol 2004b)

$$C_{ma,i} = C_{0,\beta} + \varphi \left[\Delta \beta_i \left(t_i\right) \right]^{\lambda} \qquad (4.6)$$

where $C_{0,\beta}$ is the fixed cost for maintenance, which is not affected by $\Delta\beta_i(t_i)$. φ and λ are the cost parameters to represent the relation between the maintenance cost and reliability index improvement. Figure 4.7(a) shows the effect of λ on the maintenance cost $C_{ma,i}$. The reliability index profile associated with the reliability improvements caused by two maintenance interventions at times $t_{ma,1}$ and $t_{ma,2}$ is illustrated in Figure 4.7(b). According to Eq. (4.6), the first maintenance needs less cost than the second maintenance (i.e., $C_{ma,1} < C_{ma,2}$), since the reliability improvement after the first maintenance is less than the improvement after the second maintenance (i.e., $\Delta\beta_1 < \Delta\beta_2$).

The maintenance intervention can reduce the deterioration rate of the reliability index profile. Accordingly, the maintenance cost $C_{ma,i}$ is estimated as (Kong and Frangopol 2004b)

$$C_{ma,i} = C_{0,\beta} + \eta \left[\Delta r_{\beta,i}\left(t_i\right) \right]^{\rho} \qquad (4.7)$$

where the fixed maintenance cost $C_{0,\beta}$ is independent of the reliability index deterioration rate reduction $\Delta r_{\beta,i}(t_i)$. η and ρ are the cost parameters to represent the relation between $C_{ma,i}$ and $\Delta r_{\beta,i}(t_i)$. The relations between $C_{ma,i}$ and $\Delta r_{\beta,i}(t_i)$ for $\rho < 1$, $\rho = 1$ and $\rho > 1$ are provided in Figure 4.8(a). Figure 4.8(b) shows the reliability profile, which is based on the deterioration rate reduction $\Delta r_{\beta,i}(t_i)$ caused by two maintenance interventions. The reduction of deterioration rate $\Delta r_{\beta,}$ after the first maintenance is less than Δr_{β} after the second one (i.e., $\Delta r_{\beta,1} < \Delta r_{\beta,2}$). Therefore, the cost for the first maintenance is less than the cost for the second maintenance (i.e., $C_{ma,1} < C_{ma,2}$) as indicated in Eq. (4.7).

Furthermore, the maintenance cost may depend on the reliability level before applying the maintenance intervention. For example, an improvement of β from 4 to 5 could require more maintenance cost than an improvement of β from 2 to 3, even though the amount of $\Delta\beta$ is equal to 1.0. Therefore, the maintenance cost $C_{ma,i}$ is obtained as (Kong and Frangopol 2004b)

$$C_{ma,i} = C_{0,\beta} + \varepsilon_i \left[\varphi \left\{ \Delta\beta_i \left(t_i\right) \right\}^{\lambda} \right] \qquad (4.8)$$

where ε_i is the multiplier representing the effect of the reliability level before a maintenance application on the maintenance cost $C_{ma,i}$. The multiplier ε_i is formulated as (Kong and Frangopol 2004b)

$$\varepsilon_i = \varepsilon_0 + \omega \left[\beta(t_i) \right]^{\psi} \qquad (4.9)$$

in which ε_0 is the constant multiplier, and ω and ψ are the parameters used to formulate the multiplier ε_i. Figure 4.9(a) illustrates the multiplier ε_i of Eq. (4.9) for $\psi < 1$, $\psi = 1$ and $\psi > 1$. The effects of the parameters ψ and ε_0 on the relation between the reliability index at time horizon (i.e., $\beta(t_h)$, where the time horizon $t_h = 40$ years) and the normalized cost of the first maintenance intervention (i.e., $C_{ma,1}/C_{ma}$, where $C_{ma,1}$ is the cost of the first maintenance and C_{ma} is the total maintenance cost) are shown in Figures 4.9(b) and 4.9(c), respectively. $\beta(t_h)$ decreases with an increase in $C_{ma,1}/C_{ma}$, because the reliability index at the first time of maintenance is usually higher than that associated with the second time of maintenance, and more cost is allocated with later maintenance interventions.

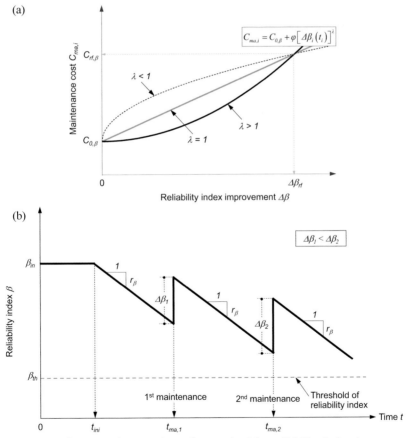

Figure 4.7 Maintenance interventions characterized by reliability index improvement: (a) maintenance cost; and (b) reliability index profile (adapted from Kong and Frangopol 2004b).

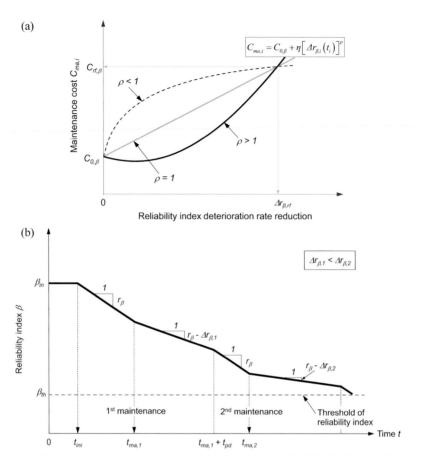

Figure 4.8 Maintenance interventions characterized by reliability index deterioration rate reduction: (a) maintenance cost; and (b) reliability index profile (adapted from Kong and Frangopol 2004b).

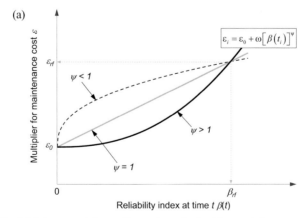

Figure 4.9 Maintenance interventions characterized by maintenance cost multiplier: (a) maintenance cost.

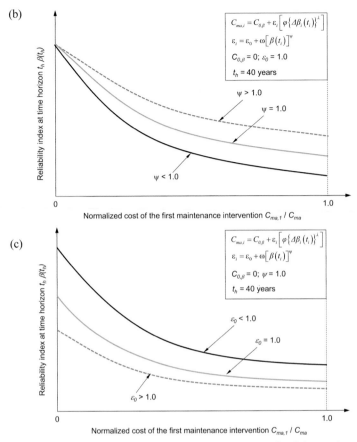

Figure 4.9*(Contd.)* Maintenance interventions characterized by maintenance cost multiplier: (b); reliability index at time horizon affected by the parameter ψ; and (c) reliability index at time horizon affected by the parameter ε_0 (adapted from Kong and Frangopol 2004b).

4.3.2   Maintenance Modeling Based on Lifetime Functions

The lifetime functions refer to the distribution functions to estimate the probability that a structure survives at any time t [i.e., survival function $S(t)$ of Eq. (1.16)], instantaneous rate of failure at time t [i.e., hazard function $h(t)$ of Eq. (1.18)], cumulative risk until time t [i.e., cumulative hazard function $H(t)$ of Eq. (1.19)], and mean residual life of a structure surviving until time t (i.e., mean residual life function $L(t)$ of Eq. (1.20)) as described in Section 1.3.3. The effect of PM and EM on the bridge service life and bridge performances of components and system can be modeled based on the lifetime functions. If the proactive PMs are applied to non-deteriorated bridge components at a uniform time interval $t_{pm,d}$, the deterioration initiation time can be delayed. Assuming the delayed time is half of $t_{pm,d}$, the deterioration initiation time $t_{ini,i}$ after the ith proactive PM is (Yang et al. 2006a, 2006b)

$$t_{ini,i} = t_{ini} + i \cdot \frac{t_{pm,d}}{2} \qquad (4.10)$$

where t_{ini} = deterioration initiation time without maintenance intervention. Therefore, the survival function $S(t)$ associated with the proactive PMs is expressed as

$$S(t) = S_0(t) \qquad \text{for } t < t_{ini,i} \qquad (4.11a)$$

$$S(t) = S_0(t - t_{ini,i}) \qquad \text{for } t \geq t_{ini,i} \qquad (4.11b)$$

where $S_0(\cdot)$ is the survival function without maintenance. Figure 4.10 shows the comparison among the probabilities of survival $S(t)$ (or reliabilities $p_s(t)$) of components and system with and without maintenances (i.e., proactive and reactive PMs, and EM). The series-parallel system used in Figure 4.10 consists of the components 1 and 2 connected in series and component 3 connected in parallel. The same survival function is used for all components (i.e., $S_1(t) = S_2(t) = S_3(t) = S(t)$, where $S_i(t)$ is the survival function of the *i*th component). Assuming that the safety margins of the components 1, 2 and 3 are independent, the system reliability $S_{sys}(t)$ is computed as

$$S_{sys}(t) = 1 - [1 - S(t)] \cdot [1 - S(t)^2] \qquad (4.12)$$

The survival function of the three components without maintenance is assumed as

$$S_0(t) = \exp[-(0.01 \cdot t)^2] \qquad (4.13)$$

Figure 4.10(a) shows the reliabilities under the proactive PMs applied to the components at the uniform time interval between proactive PMs $t_{pm,d}$ = 3 years. The deterioration initiation time without maintenance intervention t_{ini} of the components is 15 years. Figure 4.10(a) indicates that the proactive PM can delay the deterioration initiation time and reduce the decrease in reliability for the components and system.

If the reactive PM is applied at uniform time intervals $t_{pm,d}$ after the deterioration initiation time t_{ini}, the survival function $S(t)$ after the *i*th reactive PM is expressed as (Kececioglu 1995)

$$S(t) = \left[S_0 \left(t_{pm,d} \right) \right]^i S_0 \left(t - i \cdot t_{pm,d} \right) \qquad (4.14)$$

The time interval $t - i \cdot t_{pm,d}$ in Eq. (4.14) indicates the time since the *i*th reactive PM. Figure 4.10(b) shows the reliabilities of components and system with and without reactive maintenance. Herein, the deterioration initiation time t_{ini} is assumed as zero. The reactive PMs at time t = 30 years and 60 years reduce the deterioration rate of a component and system to its initial rate (i.e., slope of a reliability after the reactive PM becomes the initial slope at time t = 0).

The EM is applied when the reliability of the component reaches a predefined threshold. The EM for a bridge component can restore its performance to the initial value at time t = 0 by replacing the existing component with a new one. Accordingly, $S(t)$ for the *i*th EM is expressed as (Yang et al. 2006a, 2006b)

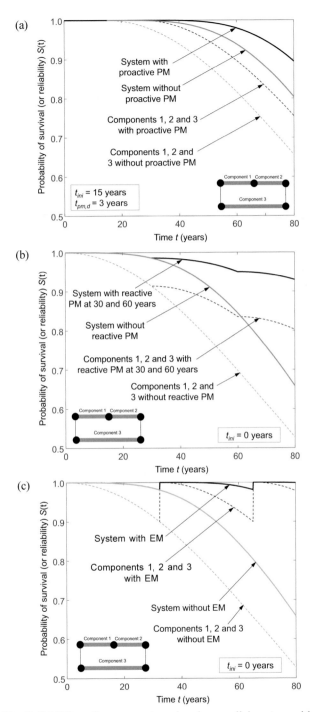

Figure 4.10 Reliabilities of components and series-parallel system with maintenances: (a) proactive PM; (b) reactive PM; and (c) EM.

$$S(t) = S_0(t) \qquad\qquad \text{for } t < t_{em,1} \qquad\qquad (4.15a)$$

$$S(t) = S_0(t - t_{em,i}) \qquad\qquad \text{for } t \geq t_{em,i} \qquad\qquad (4.15b)$$

where $t_{em,i}$ = time for the ith EM application; and $S_0(\cdot)$ = survival function without maintenance. Figure 4.10(c) provides the reliabilities of components and system with and without EM, where the threshold of the reliability is 0.90 for all components, and the deterioration initiation time t_{ini} is zero.

The total maintenance cost can be computed considering the times and types of maintenance applied on bridge components. The present value of the expected total maintenance cost C_{ma} is (Yang et al. 2006a, 2006b; Okasha and Frangopol 2010a, 2010b).

$$C_{ma} = \sum_{i=1}^{N_{ma}} \left[\frac{C_{ma,i}}{\left(1+r_{dis}\right)^{t_{ma,i}}} \right] \qquad\qquad (4.16)$$

where $C_{ma,i}$ = ith maintenance cost among N_{ma} maintenances; $t_{ma,i}$ = application time of the ith maintenance; and r_{dis} = discount rate of money. The total number of maintenances N_{ma} is the sum of the numbers of PMs and EMs. The maintenance cost $C_{ma,i}$ depends on the type of maintenance applied.

4.3.3   Maintenance Modeling Based on Damage Propagation

The effects of maintenance on structural performance and cost can be modeled based on damage propagation and resistance degradation under uncertainty (Mori and Ellingwood 1994a; Frangopol et al. 1997b; Enright and Frangopol 1999b; Estes and Frangopol 2001). In general, a maintained structure cannot recover its initial condition and performance, since all damages may not be perfectly detected due to the uncertainty associated with damage detection. Accordingly, the degree of damage d_{ma} after the application of a maintenance at time t can be estimated as (Frangopol et al. 1997b)

$$d_{ma} = d(t) \qquad\qquad \text{for} \qquad 0 \leq d(t) < d_{min} \qquad\qquad (4.17a)$$

$$d_{ma} = \frac{d_{min} + d(t)}{2} \qquad\qquad \text{for} \qquad d_{min} \leq d(t) < d_{max} \qquad\qquad (4.17b)$$

$$d_{ma} = d_{0.5} \qquad\qquad \text{for} \qquad d_{max} \leq d(t) \qquad\qquad (4.17c)$$

where $d(t)$ = degree of damage at time t; d_{min} = minimum detectable degree of damage; d_{max} = degree of damage associated with probability of damage detection equal to 1.0; and $d_{0.5}$ = degree of damage associated with probability of damage detection equal to 0.5. Considering the effect of both aging and corrosion deterioration, the expected moment capacity $E(M_{ma})$ of an RC bridge girder after maintenance is expressed as (Frangopol et al. 1997b)

$$E(M_{ma}) = M_{ma} \qquad\qquad \text{for} \qquad M_{ag}(t) \geq M_{ma} \qquad\qquad (4.18a)$$

$$E(M_{ma}) = \frac{M_{ma} + M_{ag}(t)}{2} \qquad \text{for} \qquad M_{ag}(t) < M_{ma} \qquad (4.18b)$$

where M_{ma} is the moment capacity after maintenance to remove corrosion damage, and $M_{ag}(t)$ is the moment capacity at time t under age deterioration only. If the deterioration due to aging does not affect structural performance seriously (i.e., $M_{ag}(t) \geq M_{ma}$), the expected moment capacity $E(M_{ma})$ is equal to M_{ma} (see Eq. (4.18a)). Otherwise, the expected moment capacity $E(M_{ma})$ is the mean of $M_{ag}(t)$ and M_{ma} (see Eq. (4.18b)). Furthermore, based on the relation between moment capacities before and after maintenance (i.e., denoted as M_{ma} and M_{be}, respectively), the maintenance cost is expressed as (Mori and Ellingwood 1994b, Frangopol et al. 1997b)

$$C_{ma} = \alpha_{ma} \left(\frac{M_{ma} - M_{be}}{M_0} \right)^{\gamma} \qquad (4.19)$$

where α_{ma} is the maintenance cost, γ is the model parameter, and M_0 is the initial moment capacity of a bridge girder. The effects of maintenance on degree of damage and moment capacity (see Eqs. (4.17) and (4.18), respectively) and maintenance cost (see Eq. (4.19)) are used to compute the probability of failure after maintenance, and formulate the expected total life-cycle cost.

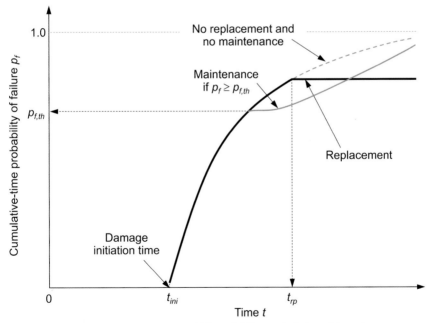

Figure 4.11 Cumulative-time probability of failure of girder with maintenance or replacement.

Using the resistance degradation function, the effects of PM, EM, and RP on the probability of failure and remaining service life of a bridge system have been investigated by Enright and Frangopol (1999b). As an illustrative example, an existing RC T-beam highway bridge (i.e., Highway Bridge L-18-BG located in Pueblo, Colorado) is used. It is assumed that the cross-sectional area of the reinforcement is reduced due to corrosion. The corrosion initiation and propagation prediction are used to formulate the resistance degradation function. The time-dependent probability of failure of both bridge components and system is computed using the resistance degradation function (Enright 1998). The types and times of maintenance affect the probability of failure and remaining service life of a bridge system. When maintenance and replacement are applied, the cumulative-time failure probability (i.e., probability of failure over the time period) of a girder is shown in Figure 4.11. Maintenance is performed if the cumulative-time failure probability reaches the threshold $p_{f,th}$, and a girder is replaced at time t_{rp}.

The maintenance modeling for fatigue-sensitive steel bridges was investigated by Liu et al. (2010b). Through the fatigue reliability analysis after retrofitting, the retrofit area of a fatigue critical location is determined, where the rectangular cut-off retrofitting is applied to remove tensile stress concentrations as shown in Figure 4.12. The reliability of the fatigue critical location after retrofitting is assessed and predicted by using finite element (FE) analysis with the monitored data. In order to optimize the retrofit area, bi-objective optimization with maximizing reliability and minimizing the retrofit area is adopted.

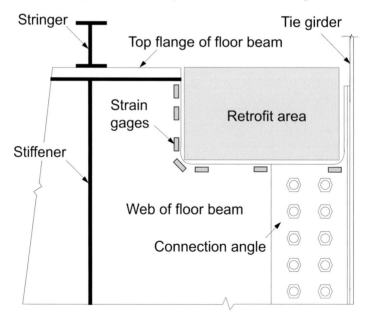

Figure 4.12   Retrofit of a fatigue critical location of a steel bridge (adapted from Liu et al. 2010b).

4.3.4 Maintenance Modeling Based on Event Tree

The effects of maintenance actions on bridge performance, service life extension and maintenance cost can be estimated using the event tree (Frangopol et al. 1997b; Kim et al., 2013). The event tree consists of the branches representing all the possible events. Each branch has its outcome and occurrence probability. The uncertainties associated with the damage propagation, inspection and maintenance are considered in the occurrence probability. The extended service life $t_{life,ex}$ after maintenance is formulated with the event tree, as indicated in Eq. (2.29). Using the PDF $f(t)$ and cumulative distribution function (CDF) $F(t)$ of the service life shown in Figure 4.13, the reliability $p_s(t_0)$ and probability of failure $p_f(t_0)$ at time t_0 (i.e., probability that the service life T is less than t_0) can be computed as (Leemis 2009; Soliman et al. 2016)

$$p_f(t_0) = P(t_0 > T) = \int_0^{t_0} f(t)dt = F(t_0) \qquad (4.20a)$$

$$p_s(t_0) = P(t_0 \le T) = \int_{t_0}^{\infty} f(t)dt = 1 - F(t_0) \qquad (4.20b)$$

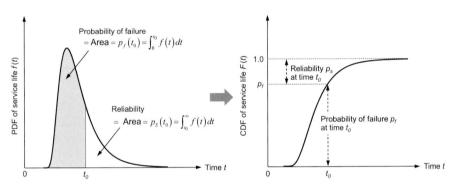

Figure 4.13 PDF and CDF of service life *T*.

As shown in Eq. (4.20a) and Figure 4.13, the probability of failure $p_f(t_0)$ is the probability that the service life T is less than time t_0. $p_f(t_0)$ corresponds to the area of the PDF of service life before time t_0 [i.e., $p_f(t_0) = F(t_0)$]. The reliability $p_s(t_0)$ is equal to $1 - F(t_0)$. Figure 4.14 presents the CDFs of the initial service life $t_{life,0}$ and the extended service life $t_{life,ex}$ of a deteriorating structure. The initial service life $t_{life,0}$ is lognormally distributed with the mean of 25.27 years and standard deviation of 4.18 years (denoted as LN(25.27; 4.18)). The service life is extended by applying the maintenance at 20 years and 30 years. It is assumed that the probability of damage detection is one, and two types of maintenance result in the time delay of deterioration t_d. The time delay t_d of the first maintenance is represented by LN(10; 2). The second maintenance is associated with the time delay of deterioration t_d defined as LN(20; 4). Detailed information can be found in Kim et al. (2013). Using Eq. (4.20) with the CDFs in Figure 4.14, the reliability and probability of failure at a specific time can be estimated.

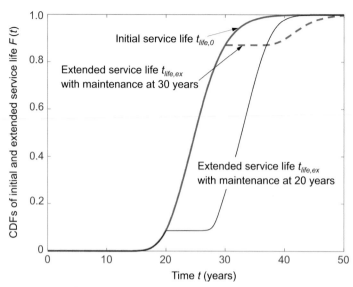

Figure 4.14 CDFs of initial and extended service life when maintenance is applied at 20 years or 30 years.

Furthermore, the expected total inspection and maintenance cost C_{ma} is expressed as

$$C_{ma} = \sum_{i=1}^{n_b} C_{ma,i} \cdot P_{b,i} \qquad (4.21)$$

where n_b = number of branches in the event tree; $C_{ma,i}$ = inspection and maintenance cost for the ith branch; and $P_{b,i}$ = occurrence probability of the ith branch. For example, the event tree in Figure 4.15, which is based on a single inspection at time t_{insp} and two maintenance types (i.e., Maintenances A and B), can be used to formulate the reliability, expected extended service life and costs of inspection and maintenance. Table 4.2 provides the outcomes (i.e., extended service life and inspection and maintenance cost) and occurrence probabilities associated with the branches of the event tree in Figure 4.15. Maintenance A is applied when the identified degree of damage is between d_A and d_B. Maintenance B is applied when the identified degree of damage is larger than d_B. The expected inspection and maintenance cost C_{ma} associated with the event tree of Figure 4.15 is

$$
\begin{aligned}
C_{ma} = {} & \left(C_{ins,s} + C_{ins,d}\right) \times P\left(t_{insp} \leq t_{life,0}\right) \times P_{det} \times P\left(0 \leq d < d_A\right) \\
& + \left(C_{ins,s} + C_{ins,d} + C_{ma,A}\right) \times P\left(t_{insp} \leq t_{life,0}\right) \times P_{det} \times P\left(d_A \leq d < d_B\right) \\
& + \left(C_{ins,s} + C_{ins,d} + C_{ma,B}\right) \times P\left(t_{insp} \leq t_{life,0}\right) \times P_{det} \times P\left(d_B \leq d\right) \\
& + C_{ins,s} \times P\left(t_{insp} \leq t_{life,0}\right) \times \left(1 - P_{det}\right) + C_{ins,s} \times P\left(t_{insp} > t_{life,0}\right)
\end{aligned}
\qquad (4.22)
$$

where $C_{ins,s}$ and $C_{ins,d}$ are the costs for scheduled and in-depth inspections, and $C_{ma,A}$ and $C_{ma,B}$ indicate the costs for maintenances A and B, respectively. The occurrence probability of Branch 4 integrates three events: (a) scheduled inspection is performed before the initial service life (i.e., $t_{insp} \leq t_{life,0}$); (b) scheduled inspection detects the damage; (c) in-depth inspection identifies that the degree of the detected damage ranges between d_A and d_B. Therefore, the occurrence probability of Branch 4 becomes $P(t_{insp} \leq t_{life,0}) \times P_{det} \times P(d_A \leq d < d_B)$ and the associated cost for inspection and maintenance is $C_{ins,s} + C_{ins,d} + C_{ma,A}$ as shown in Table 4.2.

Table 4.2 Outcomes and occurrence probabilities associated with branches in Figure 4.15

	Extended service life	Inspection and maintenance cost	Occurrence probability
Branch 1	$t_{life,0}$	$C_{ins,s}$	$P(t_{insp} > t_{life,0})$
Branch 2	$t_{life,0}$	$C_{ins,s}$	$P(t_{insp} \leq t_{life,0})$ $\times (1 - P_{det})$
Branch 3	$t_{life,0}$	$C_{ins,s} + C_{ins,d}$	$P(t_{insp} \leq t_{life,0}) \times P_{det}$ $\times P(0 \leq d < d_A)$
Branch 4	$t_{life,A}$	$C_{ins,s} + C_{ins,d} + C_{ma,A}$	$P(t_{insp} \leq t_{life,0}) \times P_{det}$ $\times P(d_A \leq d < d_B)$
Branch 5	$t_{life,B}$	$C_{ins,s} + C_{ins,d} + C_{ma,B}$	$P(t_{insp} \leq t_{life,0}) \times P_{det}$ $\times P(d_B \leq d)$

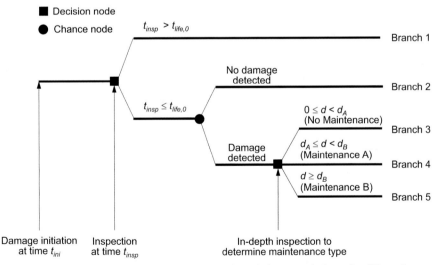

Figure 4.15 Event tree for the formulation of the extended service life and maintenance cost under uncertainty.

The probability of applying maintenance P_{ma} may depend on the degree of detectability of the damage, and bridge managers' willingness for maintenance application (Estes and Frangopol 2001). The availability of financial resources for maintenance can affect the bridge managers' willingness for maintenance application. Estes and Frangopol (2001) used the probability of applying maintenance P_{ma} for optimum maintenance planning of RC bridge decks, where four approaches (i.e., delayed, proactive, linear and idealized) to represent P_{ma} were considered. The probabilities of applying maintenance associated with these four approaches are illustrated in Figure 4.16, where d_{min} and d_{max} are the minimum and maximum detectable degrees of damage, respectively. The probability P_{ma} can be implemented in the event tree for the formulation of the extended service life and expected inspection and maintenance cost (Kim et al. 2011).

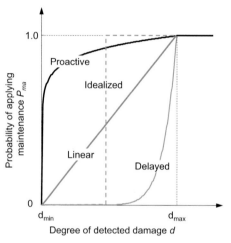

Figure 4.16   Probability of applying maintenance.

4.4   CORRELATION OF SERVICE LIFE EXTENSIONS DUE TO MAINTENANCE ACTIONS

During the service lives of existing bridges, maintenance actions are generally applied to multiple bridge components at the same time. The same maintenance types and work processes are likely to be applied for the same type of detected damage. Therefore, the maintenance actions applied to a deteriorating bridge can be correlated. When the service life extension caused by maintenance is treated as a random variable, the correlation coefficient between the service life extensions due to the ith and jth maintenances ρ_{ij} is defined as

$$\rho_{ij} = \frac{E\left[\left\{t_{ex,i} - E\left[t_{ex,i}\right]\right\} \cdot \left\{t_{ex,j} - E\left[t_{ex,j}\right]\right\}\right]}{\sigma_i \cdot \sigma_j} \tag{4.23}$$

where $t_{ex,i}$ is the service life extension by applying the ith maintenance, and $E[t_{ex,i}]$ and σ_i are the mean and standard deviation of the service life extension due to the ith maintenance, respectively.

4.4.1 Effect of Correlation among Service Life Extensions on Component Service Life

In order to investigate the effect of the correlation among service life extensions of components due to various maintenance actions on the extended service life of components, three components are considered. Component 1 has the initial service life $t_{life,0}$ represented by a lognormal distribution with a mean of 30 years and a standard deviation of 6 years (denoted as LN(30; 6)). The EMs for component 1 are applied at 30 years, 60 years and 90 years. The initial service life and EM application times of components 2 and 3 are provided in Table 4.3. In this example, it is assumed that the service life extension through EM is the same as the initial service life (i.e., $t_{ex,i} = t_{life,0}$). The extended service life $t_{life,ex}$ after each EM is computed as

$$t_{life,ex} = t_{em,i} + t_{life,0} \qquad \text{for } t'_{life,ex} \geq t_{em,i} \qquad (4.24a)$$

$$t_{life,ex} = t'_{life,ex} \qquad \text{for } t'_{life,ex} < t_{em,i} \qquad (4.24b)$$

where $t'_{life,ex}$ is the service life before applying EM, and $t_{em,i}$ is the ith EM application time. It should be noted that EM before reaching the service life (i.e., $t'_{life,ex} \geq t_{em,i}$) results in the service life extension as indicated in Eq. (4.24a). The correlation matrix of the service life extensions by EMs $\rho_{(k)}$ for the kth component is expressed as

$$\boldsymbol{\rho}_{(k)} = \begin{bmatrix} \rho_{11} & \rho_{12} & \rho_{13} \\ \rho_{21} & \rho_{22} & \rho_{23} \\ \rho_{31} & \rho_{32} & \rho_{33} \end{bmatrix} = \begin{bmatrix} 1 & 0 & 0 \\ 0 & 1 & 0 \\ 0 & 0 & 1 \end{bmatrix} = \mathbf{I} \qquad \begin{array}{l} \text{for independent} \\ \text{service life} \\ \text{extensions by EMs} \end{array} \qquad (4.25a)$$

$$\boldsymbol{\rho}_{(k)} = \begin{bmatrix} \rho_{11} & \rho_{12} & \rho_{13} \\ \rho_{21} & \rho_{22} & \rho_{23} \\ \rho_{31} & \rho_{32} & \rho_{33} \end{bmatrix} = \begin{bmatrix} 1 & 1 & 1 \\ 1 & 1 & 1 \\ 1 & 1 & 1 \end{bmatrix} = \mathbf{J} \qquad \begin{array}{l} \text{for perfectly positive} \\ \text{correlated service life} \\ \text{extensions by EMs} \end{array} \qquad (4.25b)$$

where ρ_{ij} is the correlation coefficient between service life extensions due to the ith and jth EMs (see Eq. (4.23)), $\mathbf{I}$ is the identity matrix, and $\mathbf{J}$ is the matrix of ones. The number of rows (or columns) of the correlation matrix $\rho_{(k)}$ corresponds to the number of EMs applied to the kth component.

Table 4.3 Initial service life of three components and times of application of essential maintenances

	Initial service life $t_{life,0}$ (years)	Essential maintenance (EM) application times t_{em} (years)
Component 1	LN(30; 6)	{30, 60, 90}
Component 2	LN(20; 4)	{20, 40, 60}
Component 3	LN(10; 2)	{10, 20, 30}

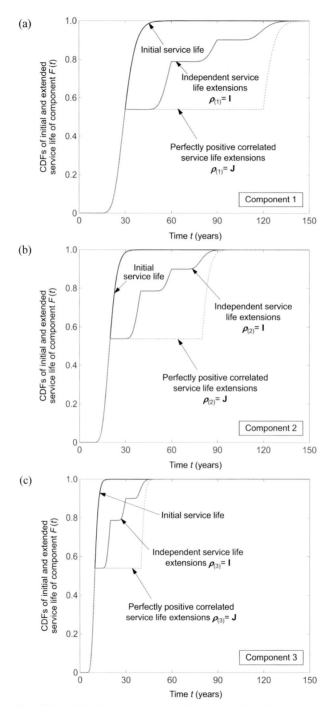

Figure 4.17 CDFs of initial and extended service life of three components with independent and perfectly positive correlated service life extensions: (a) component 1; (b) component 2; and (c) component 3.

When the service life extensions by EMs are independent (i.e., $\boldsymbol{\rho}_{(k)} = \mathbf{I}$) and perfectly positive correlated (i.e., $\boldsymbol{\rho}_{(k)} = \mathbf{J}$), the CDFs of initial service life $t_{life,0}$ and extended service life $t_{life,ex}$ of components 1, 2 and 3 are illustrated in Figures 4.17(a), 4.17(b) and 4.17(c), respectively. The CDF of the service life $F(t)$ in Figure 4.17 indicates the probability that the service life is less than time t (or probability of failure at time t). Therefore, the reliability p_s and probability of failure p_f of each component can be estimated with the CDFs in Figure 4.17. From the comparison among the CDFs in Figure 4.17, it can be seen that the case of perfectly positive correlated service life extensions (i.e., $\boldsymbol{\rho}_{(k)} = \mathbf{J}$) leads to the minimum probability of failure for components 1, 2 and 3, and the probability of failure p_f of component 3 is the largest among the probabilities of failure of all components.

4.4.2 Effect of Correlation among Service Life Extensions on System Service Life

The effect of the correlation among service life extensions by applying EMs on the extended service life of systems is investigated with series, parallel and series-parallel systems with component 1 in series. These systems consist of components 1, 2 and 3 defined in Table 4.3. The correlation matrix of the service life extensions $\boldsymbol{\rho}_{(sys)}$ for the system with EMs is expressed as

$$\boldsymbol{\rho}_{(sys)} = \mathbf{I} \qquad \text{for independent service life extension by EMs} \qquad (4.26a)$$

$$\boldsymbol{\rho}_{(sys)} = \mathbf{J} \qquad \text{for perfectly positive correlated service life extension by EMs} \qquad (4.26b)$$

where $\mathbf{I}$ is a 9×9 matrix whose diagonal elements are one and all off-diagonal elements are zero, and $\mathbf{J}$ is a 9×9 matrix whose all elements are one. The number of rows (or columns) of the correlation matrix $\boldsymbol{\rho}_{(sys)}$ in Eq. (4.26) is equal to the product of the number of components and the number of EMs to be applied (i.e., $3 \times 3 = 9$).

Figures 4.18, 4.19 and 4.20 show the CDFs of initial service life $t_{life,0}$ and extended service life $t_{life,ex}$ for series, parallel and series-parallel systems with component 1 in series, respectively, where the independent and perfectly positive correlated service extensions by applying EMs (i.e., $\boldsymbol{\rho}_{(sys)} = \mathbf{I}$ and $\boldsymbol{\rho}_{(sys)} = \mathbf{J}$, respectively) are considered. The CDFs in Figures 4.18, 4.19 and 4.20 correspond to the system probability of failure p_f at time t, which are obtained using Eqs. (1.13a), (1.13b) and (1.13c) with the CDFs in Figures 4.17(a), 4.17(b) and 4.17(c). When the safety margins of the three components are independent, the CDFs of $t_{life,0}$ and $t_{life,ex}$ associated with $\boldsymbol{\rho}_{(sys)} = \mathbf{I}$ and $\boldsymbol{\rho}_{(sys)} = \mathbf{J}$ are compared in Figure 4.18(a) for the series system, in Figure 4.19(a) for the parallel system, and in Figure 4.20(a) for the series-parallel system. Figures 4.18(b), 4.19(b) and 4.20(b) provide the CDFs of $t_{life,0}$ and $t_{life,ex}$ for the series, parallel, series-parallel systems, respectively, when the safety margins of three components are perfectly positive correlated. The comparison of the CDFs of $t_{life,0}$ and $t_{life,ex}$ in Figures 4.18, 4.19 and 4.20 shows that:

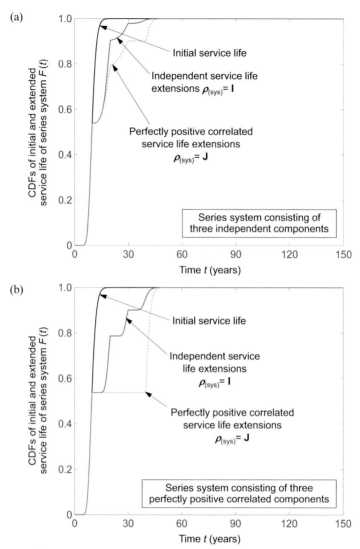

Figure 4.18   CDFs of initial and extended service life for series system with independent and perfectly positive correlated service life extensions: (a) independent components; and (b) perfectly positive correlated components.

- The series, parallel and series-parallel systems without EM result in larger system probability of failure p_f than that of the same systems with EM.
- The system probability of failure p_f associated with perfectly positive correlated service life extensions is less than that associated with independent service life extensions.

Furthermore, the CDFs of the extended service life $t_{life,ex}$ for the series, parallel and series-parallel systems are compared in Figures 4.21 and 4.22. Figures 4.21(a) and 4.21(b) show the CDFs of $t_{life,ex}$ for the systems consisting of three components

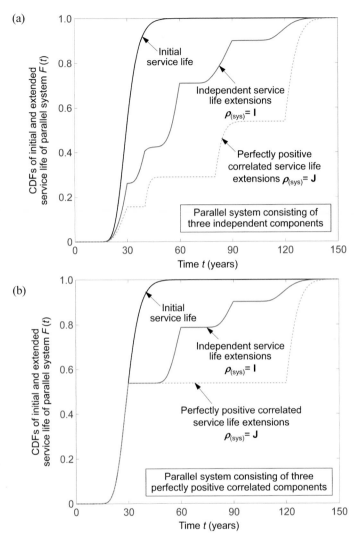

Figure 4.19   CDFs of initial and extended service life for parallel system with independent and perfectly positive correlated service life extensions: (a) independent components; and (b) perfectly positive correlated components.

with independent safety margins and independent and perfectly dependent service life extensions, respectively. The CDFs in Figure 4.22(a) and 4.22(b) are associated with the systems consisting of three components with perfectly positive correlated safety margins and independent and perfectly dependent service life extensions, respectively. From Figures 4.21 and 4.22, it can be seen that the probability of failure p_f for the series system is the largest, and the parallel system results in the smallest p_f.

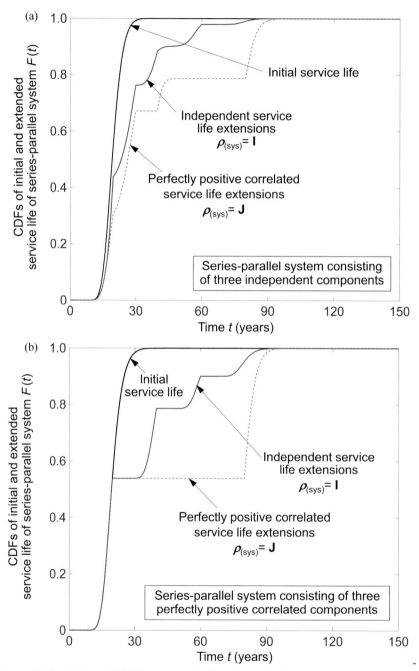

Figure 4.20 CDFs of initial and extended service life for series-parallel system with independent and perfectly positive correlated service life extensions: (a) independent components; and (b) perfectly positive correlated components.

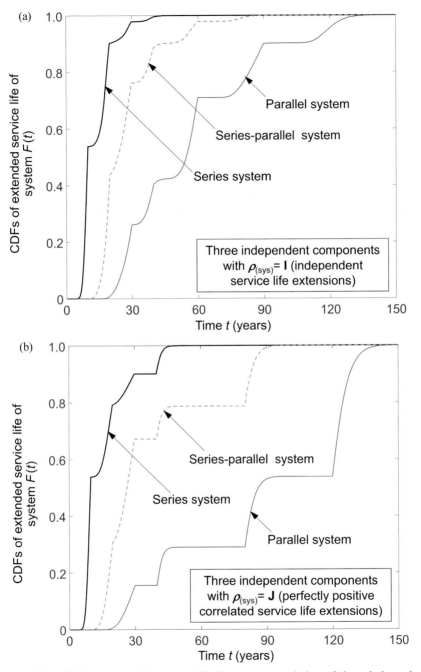

Figure 4.21 CDFs of extended service life for system consisting of three independent components: (a) independent service life extensions; and (b) perfectly positive correlated service life extensions.

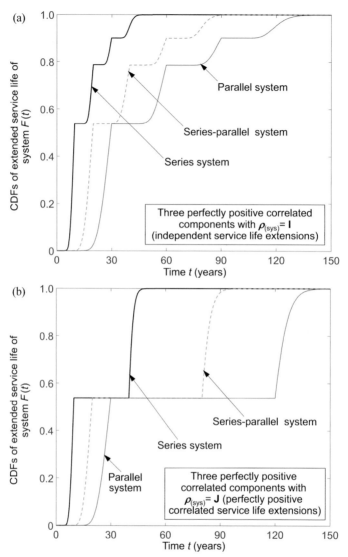

Figure 4.22 CDFs of extended service life for system consisting of three perfectly positive correlated components: (a) independent service life extensions; and (b) perfectly positive correlated service life extensions.

4.5 BRIDGE MAINTENANCE MANAGEMENT PROCESS

Bridge maintenance management is essential for bridge management agencies to maintain existing bridges safely and cost-effectively. Most developed countries and their local governments have paid significant attention to the condition and safety of deteriorating bridges, since the economic impact of deteriorating bridges is significantly high (NCHRP 2006; ASCE 2017b). Accordingly, cost-effective and practical processes for bridge maintenance management have been investigated and applied over several decades (Hurt and Schrock 2016). In general, the bridge maintenance management process consists of maintenance planning, maintenance application, and reporting and evaluating after maintenance (AASHTO 2007a).

The planning process for bridge maintenance has three phases: (a) preliminary planning, (b) budgeting, and (c) maintenance planning. Preliminary planning for bridge maintenance management is a decision making to address the requirements of agencies and users (NCHRP 2009). By integrating the needs for maintenance of the bridge network during a given time period and the information about maintenance requirements and histories, the preliminary planning should be performed. Budgeting is the process of securing financial resources to perform the maintenance plans. The effective budgeting has to be based on accurate preliminary planning with reliable cost data (NCHRP 1997; NCHRP 2003). Maintenance planning is a process to allocate the maintenance works and costs under the resources determined from the budgeting. In order to achieve maintenance planning effectively, significant investigations have been developed over several decades (Frangopol 2011; Sánchez-Silva et al. 2016). The bridge maintenance planning integrates the effects of maintenance on bridge performance improvement, service life extension, and life-cycle cost. Through the single- and multi-objective optimization process, the maintenance planning can be optimized (Frangopol and Soliman 2016; Frangopol and Kim 2019). More details about the maintenance planning for optimum life-cycle bridge management are provided in Chapter 6.

Bridge maintenances are applied when the damage is detected and identified through inspections. Decision making to determine the most appropriate maintenance types should consider the type, degree and location of the identified damage, expected performance and service life after maintenance, and available financial resources. After performing maintenance, reporting and evaluating is essential for future bridge maintenance management. The primary purposes of reporting and evaluating include making a database, sharing the information associated with bridge maintenance, and evaluating the effectiveness of the performed maintenance interventions (AASHTO 2007a). Using a bridge management system (BMS), the data for reporting and evaluating can be effectively and efficiently managed (Adey et al. 2003).

4.6 BRIDGE MANAGEMENT SYSTEMS

BMS is the integrated platform for efficient and cost-effective bridge asset management, which can be used to predict future bridge condition and maintenance needs, and to determine optimum plans and budgets (AASHTO 2007a). Many countries developed and used their own BMS to address efficiently and effectively their situations to manage deteriorating bridges. Table 4.4 provides the existing BMS used in several countries.

Table 4.4 Several existing bridge management systems
(adapted from Mirzaei et al. 2012)

Country	Owner	Name of system	First version (years)	Number of condition states
Canada	Ontario Ministry of Transportation and Stantec Consulting Ltd.	Ontario Bridge Management System (OBMS)	2002	4
Canada	Quebec Ministry of Transportation	Quebec Bridge Management System (QBMS)	2008	4
Canada	Edmonton Ministry of Transportation	Edmonton Bridge Management System (EBMS)	2006	5
Canada	Prince Edward Island Dept. of Transportation	Prince Edward Island Bridge Management System (PEI BMS)	2006	4
Denmark	Danish Road Directorate	DANBRO Bridge Management System (DANBRO)	1975	5
Finland	Finnish Transport Agency	The Finnish Bridge Management System (FBMS)	1990	4
Germany	German Federal Highway Research Institute	Bauwerk Management System (GBMS)	N/A	4
Ireland	Irish National Road Association	Eirspan	2001	5
Italy	Autonomous Province of Trento	Autonomous Province of Trento Bridge Management System (APTBMS)	2004	5
Japan	Kajima Corporation and Regional Planning Institute of Osaka	BMS@RPI (RPIBMS)	2006	5
Korea	Korean Ministry of Land, Transport and Maritime Affairs	Korea Road Maintenance Business System (KRMBS)	2003	5
Latvia	Latvian State Road Administration	Lat Brutus	2002	4
Netherlands	Dutch Ministry of transport	DISK	1985	7
Poland	Polish Railway Lines	SMOK	1997	5
Poland	Local Polish Road Administrations	SZOK	2001	5
Spain	Spanish Ministry of Public Works	SGP	2005	100
Sweden	Swedish Road Administration	Bridge and Tunnel Management System (BaTMan)	1987	3

Table 4.4 (*Contd.*)   Several existing bridge management systems
(adapted from Mirzaei et al. 2012).

Country	Owner	Name of system	First version (years)	Number of condition states
Switzerland	Swiss Federal Roads Authority	KUBA	1991	5
USA	Alabama Department of Transportation	Alabama Bridge Management System (ABMS)	1994	9
USA	American Association of State Highway and Transportation Officials	Pontis	1989	5
USA	American Association of State Highway and Transportation Officials	BRIDGIT	1985	5
Vietnam	Vietnam Ministry of transportation	Bridgeman	2001	N/A

A representative BMS in the United States is AASHTOWare Pontis, which was initially developed by Cambridge Systematics in 1989 for Federal Highway Administration (FHWA), and has been revised several times by requests of FHWA, American Association of State Highway and Transportation Officials (AASHTO), and State Departments of Transportation (FHWA 2005; Yari 2018). The main functions of AASHTOWare Pontis include (a) establishing an accurate and reliable inventory of bridge information, (b) managing inventory data and information for bridge inspection, (c) scheduling bridge inspections, and (d) managing National Bridge Inventory data and other inspection reports (AASHTO 2015). BRIDGIT is another representative BMS used in the United States (Frangopol et al. 2001). This system was developed under the AASHTO-sponsored NCHRP Program in 1985. BRIDGIT is a multi-user PC-based system, and can be operated without extensive training for users. For this reason, BRIDGIT is suitable for small bridge management agencies which may not have enough human resources for BMS operation. BRIDGIT consists of (a) inventory module, (b) inspection module, (c) maintenance, rehabilitation and replacement module, (d) analysis module, and (e) model module. Using BRIDGIT, it is possible to handle bridge inventories of thousands of bridges. Additional information on BRIDGIT is available in Hawk and Small (1998).

In Switzerland, the Swiss Federal Roads Office (FEDRO) uses KUBA software for bridge management, as indicated in Table 4.4. KUBA developed in 1991 consists of two main modules: database module (KUBA-DB) and management system (KUBA-MS). There are two additional systems: load rating system (KUBA-ST) and reporting system (KUBA-RP) (Hajdin 2008). KUBA-DB contains inventory data, inspection data, and data on completed repair projects. KUBA-MS is the set of management functions including condition assessment and prediction, specification of technically feasible actions, generation of preservation projects, optimization on project level, and establishment of working program. KUBA-DB and KUBA-MS are stand-alone systems, which can be operated separately. The data from KUBA-DB is imported to KUBA-MS for optimum maintenance programs (FHWA 2005). The theoretical backgrounds and detailed process used in KUBA-MS can be found in Hajdin (2004).

The Swedish BMS developed in 1987 is called SAFEBRO. This BMS consists of object databases, knowledge basis, and process modules. The object databases include a bridge inventory, condition data, user data, and work plans. Deterioration models, cost data, definitions of performance measures, standard repair actions, and their applications are involved in the knowledge basis. The process modules contain the user interface, data updating procedures, program analysis, optimization process, and the report generator. In 2004, SAFEBRO was replaced with BaTMan (Bridge and Tunnel Management), which is a Web-based management system for both bridges and tunnels. Additional information on BaTMan including the associated theoretical background is available in SRA (1996a, 1996b, 2002), FHWA (2005), Hallberg and Racutanu (2007) and Safi et al. (2013).

The existing German BMS contains the condition assessment and optimization procedures on object and network level (Haardt 2002; Haardt and Holst 2008). This BMS has four interconnected modules: (a) BMS-MV (measure variants), (b) BMS-MB (measure evaluation), (c) BMS-EP (maintenance program), and (d) BMS-SB (scenario building). BMS-MV provides all the information for subsequent computing process of BMS. BMS-MB evaluates maintenance alternatives on object level, in which cost-benefit analysis for maintenance alternatives proposed by BMS-MV is performed. BMS-EP optimizes maintenance planning on the network level and provides maintenance interventions. BMS-SB evaluates maintenance strategies for objects on the network level (Haardt 2002; FHWA 2005).

Further descriptions on BMS of other countries can be found in FHWA (2005), Mirzaei et al (2012) and Scutaru et al. (2018).

4.7 CONCLUDING REMARKS

In this chapter, the bridge maintenance types and their effects on performance, service life, service life extension and cost are described. The primary characteristics of PM, EM and RP are presented. The effects of PM, EM and RP on bridge components and systems are compared. In order to quantify these effects rationally, maintenance modeling is essential. The maintenance models based on performance profile, lifetime function, damage propagation and event tree are presented. The maintenance model based on the performance profile considers the relations among reliability index, service life and cost. The effects of PM and EM and RP on the delay of deterioration initiation time, reduction of reliability deterioration rate and improvement of reliability are addressed in the formulation of the survival function. The effects of PM, EM and RP on structural performance and cost are modeled based on damage propagation and resistance degradation under uncertainty. The event tree for maintenance modeling is formulated with the branches representing all the possible events by taking into account the uncertainties associated with predictions of service life, damage detection, maintenance intervention and maintenance cost. The effect of correlation among the service life extensions due to various maintenances on the extended service life of components and systems is investigated. The general process for bridge maintenance, which consists of maintenance planning, maintenance application, and reporting and evaluating after maintenance, and BMS are provided.

Chapter 5

Life-Cycle Performance Analysis and Optimization

NOTATIONS (continued)	
t_{insp}	= inspection application time
$t_{life,d}$	= predefined lifetime
$t_{life,ex}$	= expected extended service life
$t_{life,0}$	= initial service life
t^*_{pm}	= optimum preventive maintenance application time
t_{mdl}	= maintenance delay
r_{dis}	= annual discount rate of money
$VI(t)$	= time-dependent vulnerability index
β_{min}	= minimum reliability index during lifetime
β_{th}	= predefined target reliability
ρ_{ij}	= coefficient of correlation between the values of the ith and jth objectives
$\rho_p(f_i, f_j)$	= Pearson's coefficient of correlation between the two objectives of minimizing f_i and f_j
$\rho_s(f_i, f_j)$	= Spearman's coefficient of correlation between the two objectives of minimizing f_i and f_j
$\Omega^{(-)}$	= objective set consisting of the objectives to be minimized
$\Omega^{(+)}$	= objective set consisting of the objectives to be maximized

ABSTRACT

In Chapter 5, general concepts and theoretical background for optimum life-cycle bridge inspection and maintenance planning under uncertainty are presented. The optimum inspection and maintenance planning can be based on the probabilistic life-cycle performance and cost analysis. The objectives for optimum bridge inspection and maintenance planning are formulated based on time-dependent bridge performance, life-cycle cost, damage detection, service life extension, and risk. The multiple objectives can be integrated systematically into multi-objective optimization. In order to find the Pareto optimal solutions of the multi-objective optimization efficiently, the correlation among objectives, identification of essential and redundant objectives, and multi-objective optimization are investigated. The approaches for multi-attribute decision making to select the best Pareto optimal solution are presented.

5.1 INTRODUCTION

Life-cycle performance and cost analysis have been considered as effective and efficient tools for bridge design and management of existing bridges (NCHRP 2003; Frangopol and Kim 2019). Recently, bridge owners and managers encounter unfavorable circumstances such as increasing the number of deteriorating bridges, increasing demands related to efficient life-cycle bridge management, and reducing financial resources (Flintsch and Chen 2004; Kabir et al. 2014; Kim and Frangopol 2018a, 2018b). Accordingly, during the past two decades, various concepts and approaches for effective life-cycle bridge performance and cost analysis under uncertainty have been developed (Biondini and Frangopol 2016; Frangopol and Soliman 2016; Frangopol et al. 2017). Successful optimization can be achieved through appropriate prediction of time-dependent bridge performance including the estimation of effects of inspection and maintenance on bridge performance, service life and cost (Sánchez-Silva et al. 2016).

In this chapter, probabilistic concepts and approaches for optimum life-cycle bridge management planning are presented. The general procedure for life-cycle performance and cost analysis under uncertainty is described. The probabilistic life-cycle performance and cost analysis for optimum bridge design and management planning are presented. The objectives for optimum bridge inspection and maintenance planning are presented, based on time-dependent bridge performance, life-cycle cost, damage detection, service life extension, and risk. Through a multi-objective optimization process, multiple objectives can be integrated systematically for bridge inspection and maintenance planning. In order to find the Pareto front of the multi-objective optimization and to select the best Pareto optimal solution, the approaches to estimate the correlation among objectives, identifying the essential and redundant objectives, solving the multi-objective optimization, and multi-attribute decision making are provided.

5.2 GENERAL CONCEPT OF LIFE-CYCLE PERFORMANCE AND COST ANALYSIS

Life-cycle performance and cost analysis can be used for planning, design, construction, and safety and service life management of new bridges (Frangopol 2011; Frangopol et al. 2017). During the bridge planning and design process, life-cycle performance and cost analysis can provide appropriate materials to be used for the bridge components, type and size of a bridge, and construction method (Frangopol et al. 1997b; Biondini and Frangopol 2009; Frangopol 2011; Ang 2011; Soliman and Frangopol 2015). For existing bridges, optimum inspection and maintenance planning can be achieved based on probabilistic life-cycle performance and cost analysis (Kong and Frangopol 2004b; Liu and Frangopol 2005b; Biondini and Frangopol 2016).

5.2.1 General Procedure for Life-Cycle Performance and Cost Analysis under Uncertainty

Figure 5.1 shows the general procedure for life-cycle performance and cost analysis. The bridge performance prediction considers the external loading conditions and resistance deterioration mechanisms under uncertainty. In this procedure, the bridge performance deterioration is required as described in Section 2.3. The bridge performance is represented by the probabilistic structural performance indicators such as reliability, probability of failure, redundancy, vulnerability, robustness and risk as presented in Section 1.4. The effects of inspection and maintenance on the bridge performance and service life need to be addressed (see Sections 2.4, 3.2 and 4.2), where essential maintenance (EM), preventive maintenance (PM) and replacement (RP) can be taken into account. The life-cycle cost consisting of initial cost, inspection and maintenance costs, and failure cost (see Eq. (1.1)) should now be estimated. Since the bridge performance prediction and cost estimation are under uncertainty, probabilistic concepts and methods are required for life-cycle performance and cost analysis. The profiles to represent the life-cycle bridge performance and cost under uncertainty are described in Sections 1.2 and 4.3.

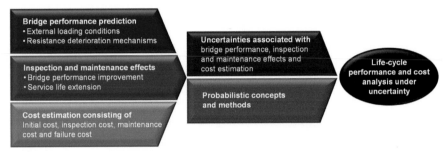

Figure 5.1   General procedure for life-cycle performance and cost analysis under uncertainty.

5.2.2   Reliability-Based Optimum Design of New Bridges

The reliability-based optimum design of a bridge component determines the geometrical properties (e.g., size of bridge components, area and spacing of reinforcement) and mechanical properties (e.g., compressive strength, tensile strength, flexural strength) to minimize the cost and/or to maximize the performance of the bridge component. The constraints for the optimization considers weight, stiffness, displacements, stresses, and buckling loads among others, which are based on the bridge design codes such as AASHTO LRFD design specification (AASHTO 2017a). The variables involved in the objective and constraints are treated as random variables (Frangopol 1985; Lin and Frangopol 1996). The reliability-based optimum design for a new bridge can be formulated as (Enevoldsen and Sørensen 1994; Al-Harthy and Frangopol 1994)

$$\text{Find } \mathbf{X} \tag{5.1a}$$

$$\text{which minimizes } C(\mathbf{X}) \tag{5.1b}$$

$$\text{subject to } g(\mathbf{X}) = \beta_{th} - \beta(\mathbf{X}) \leq 0 \tag{5.1c}$$

where $\mathbf{X}$ is a vector of design variables, β_{th} is the predefined target reliability index, $C(\mathbf{X})$ is the function represented by cost, volume and cross-sectional area, and $\beta(\mathbf{X})$ is the reliability index associated with the design variables $\mathbf{X}$. Lin and Frangopol (1996) investigated the reliability-based optimum design of RC girders, where the total material cost is minimized to find the design variables such as the area of tension reinforcement, flange width and thickness, web width and height, among others. The relation between the optimum total cost C^* and reliability index $\beta(\mathbf{X})$ of a bridge is shown in Figure 5.2(a). It is shown that an increase in $\beta(\mathbf{X})$ requires an increase in the optimum total cost C^*. Figure 5.2(b) presents the optimum total cost C^* versus the predefined target reliability index β_{th}. This figure shows that a larger β_{th} leads to a larger optimum cost C^*. The reliability-based prestressed concrete (PC) beam design is presented in Al-Harthy and Frangopol (1994), where the total area of prestressing strands is minimized under the constraints related to the reliability index of bridge components.

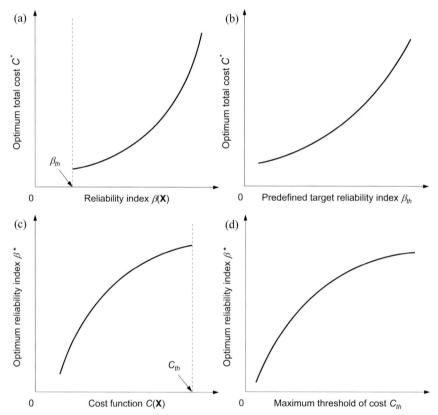

Figure 5.2 Reliability-based optimum design: (a) relation between optimum total cost and reliability index associated with the desgin; (b) relation between optimum total cost and target reliability index; (c) relation between optimum reliability index and cost; and (d) relation between optimum reliability index and maximum threshold of cost.

Instead of Eq. (5.1), the objective associated with maximizing the reliability index, and constraint expressed with the cost function can be adopted as (Fu and Frangopol 1990a, 1990b; Frangopol and Maute 2003)

Find $\mathbf{X}$ (5.2a)

which maximizes $\beta(\mathbf{X})$ (5.2b)

subject to $g(\mathbf{X}) = C(\mathbf{X}) - C_{th} \leq 0$ (5.2c) where C_{th} is the maximum threshold of cost. Figure 5.2(c) shows that a bridge associated with a larger optimum reliability index β^* needs a larger value of the cost function $C(\mathbf{X})$. The optimum reliability index β^* of a bridge increases with an increase in the maximum threshold of cost C_{th} as shown in Figure 5.2(d).

5.2.3 Life-Cycle Performance- and Cost-Based Optimum Bridge Design

The life-cycle performance- and cost-based optimum design of new bridges was developed by integrating the life-cycle performance and cost analysis into the reliability-based optimum design of bridge structures (Frangopol et al. 1997b; Frangopol and Maute 2003; Barone et al. 2014). The single-objective optimization for life-cycle optimum design can be formulated based on Eqs. (5.1) and (5.2), where the formulations of the cost function $C(\mathbf{X})$ and reliability index $\beta(\mathbf{X})$ consider the effects of inspection and maintenance on the bridge performance and life-cycle cost. Furthermore, the multi-objective optimization based on minimizing the expected life-cycle cost and/or maximizing the expected bridge performance (or the expected extended service life) can be adopted to determine the well-balanced design variables (Furuta et al. 2006; Frangopol et al. 2017). The general formulation of the multi-objective optimization of bridge design is expressed as (Deb 2001; Rao 2009; Biondini and Frangopol 2016; Arora 2017)

$$\text{Find } \mathbf{X} \tag{5.3a}$$

$$\text{which minimizes } \Omega^{(-)} = \{ f_{1(-)}(\mathbf{X}), f_{2(-)}(\mathbf{X}), ..., f_{n(-)}(\mathbf{X}) \}$$

$$\text{and/or maximizes } \Omega^{(+)} = \{ f_{1(+)}(\mathbf{X}), f_{2(+)}(\mathbf{X}), ..., f_{n(+)}(\mathbf{X}) \} \tag{5.3b}$$

$$\text{subject to } g_k(\mathbf{X}, t) \le 0 \qquad k = 1, 2, ... n_c \tag{5.3c}$$

$$\mathbf{X}^- \le \mathbf{X} \le \mathbf{X}^+ \tag{5.3d}$$

$$t_0 \le t \le t_{life,d} \tag{5.3e}$$

where $\mathbf{X}$ is the vector of design variables, $g_k(\mathbf{X}, t) \le 0$ is the kth inequality constraint among n_c time constraints, and $\mathbf{X}^-$ and $\mathbf{X}^+$ are the vectors of lower and upper bounds of the design variables. The time t should range from the initial time t_0 to predefined lifetime $t_{life,d}$. $\Omega^{(-)}$ and $\Omega^{(+)}$ are the objective sets consisting of $n(-)$ and $n(+)$ objectives to be minimized and to be maximized, respectively. Minimizing the expected life-cycle cost during the predefined lifetime $t_{life,d}$ is related to the objectives involved in $\Omega^{(-)}$. Maximizing the expected lifetime bridge performance can be an objective associated with $\Omega^{(+)}$. The design variables $\mathbf{X}$ for new bridges may include the geometrical and mechanical properties of a bridge system, and times and types of inspection and maintenance, including SHM interventions (Mori and Ellingwood 1994a, 1994b; Lin and Frangopol 1996; Frangopol et al. 1997b; Biondini and Marchiondelli 2008; Soliman and Frangopol 2015).

5.2.4   Life-Cycle Performance- and Cost-Based Optimum Management of Existing Bridges

The life-cycle performance and cost analysis can also be applied to optimize the service life management of existing bridges. The associated optimization process is based on the formulation indicated in Eq. (5.3). The comparisons between the optimization for life-cycle performance- and cost-based bridge design and management of existing bridges are summarized in Table 5.1. When a bridge is designed and constructed, its initial cost C_{ini} depends on the lifetime structural performance P_{life}, inspection and maintenance cost $C_{ins} + C_{ma}$, and failure cost C_{fail} as described in Figures 1.4 and 1.5. However, the expected life-cycle cost analysis for existing bridges consider the fixed initial cost (i.e., the initial cost of this bridge). Moreover, the formulation of the objectives considering the bridge performance prediction should be based on the time when the service life management planning of an existing bridge is initiated, and consider the inspection and maintenance interventions, including SHM, already applied. The bridge performance prediction can be updated using the information from these previous interventions. The design variables associated with existing bridges are the times and types of future inspections and maintenances for given geometrical and mechanical properties. The number of design variables for existing bridge management is less than that for bridge design.

Table 5.1   Comparison between optimization for life-cycle performance- and cost-based new bridge design and existing bridge management

	Optimization for life-cycle performance- and cost-based bridge design	Optimization for life-cycle performance- and cost-based management of existing bridges
Formulation of objectives	• Variable initial cost • Bridge performance prediction during its life-cycle (i.e., since construction to the end of service life)	• Fixed initial cost • Bridge performance prediction since the bridge management initiation • Consideration of inspection and maintenance, including SHM, interventions already applied • Updating the bridge performance and associated objective functions using information from previous interventions
Design variables	• Geometrical and mechanical properties of the new bridge • Times and types of inspections and maintenances, including SHM, during the whole service life of the bridge	• Times and types of future interventions for bridge management planning

5.3 OBJECTIVES OF OPTIMUM LIFE-CYCLE BRIDGE MANAGEMENT PLANNING

Along with increasing demands related to structural performance, safety and risk management, a significant amount of effort has been made to develop more rational and practical probabilistic concepts and approaches for bridge management (Frangopol 2011; Frangopol and Soliman 2016). Based on these developed concepts and approaches, various objectives were introduced and adopted for optimum bridge life-cycle performance- and cost-based management. The associated objectives can be categorized into: (a) performance-based objectives, (b) cost-based objectives, (c) damage detection-based objectives, (d) service life-based objectives, and (e) risk-based objectives.

5.3.1 Performance-Based Objectives

The performance-based objectives for optimum bridge life-cycle management planning are formulated by integrating the effect of maintenance interventions into the time-dependent performance indicators. The performance-based objective functions include the time-dependent reliability $p_s(t)$, reliability index $\beta(t)$, and probability of failure $p_f(t)$, where the effect of maintenance on the bridge performance can be modeled using the multi-linear and nonlinear profiles, damage propagation, and event tree as described in Section 4.3. The reliability and probability of failure at bridge system level can be extended to formulate the time-dependent redundancy index $RI(t)$ [see Eqs. (1.43) and (1.44)], vulnerability index $VI(t)$ [see Eq. (1.45)], and structural robustness $RB(t)$ [see Eq. (1.47)]. Depending on the types and times of application of inspection and maintenance actions, the bridge performance represented by these indicators varies over time. Therefore, efficient maintenance planning can be achieved through the optimization process, where the types and times of application of inspection and maintenance actions are determined.

When the optimization for bridge management is formulated, the objectives should be selected appropriately, since the bridge management planning depends on these objectives. For example, it is considered that (a) the reliability index profile is represented by a multi-linear model, (b) a single maintenance is applied before reaching the threshold of reliability index β_{th}, (c) the maintenance results in the reliability index improvement $\Delta\beta$, (d) reliability index improved after the maintenance cannot be larger than the initial reliability index β_{in}, and (e) reliability after the maintenance decreases with the initial slope r_β. Accordingly, the reliability index profile without maintenance is expressed as

$$\beta(t) = \beta_{in} \qquad\qquad \text{for } t < t_{ini} \qquad\qquad (5.4a)$$

$$\beta(t) = \beta_{in} - r_\beta (t - t_{ini}) \qquad\qquad \text{for } t \geq t_{ini} \qquad\qquad (5.4b)$$

where $\beta(t)$ is the reliability index at time t, and t_{ini} is the damage initiation time. The reliability index profile after maintenance is expressed as

$$\beta(t') = \beta_{in} \qquad\qquad \text{for } t_{ma} < t_{ini} \qquad (5.5a)$$

$$\beta(t') = \beta(t_{ma}) + \Delta\beta - r_\beta \cdot t' \qquad \text{for } t_{ma} \geq t_{ini} \qquad (5.5b)$$

where $\beta(t')$ = reliability index at time t'; and t' = time since the maintenance application time t_{ma}. The formulation and results of the optimum maintenance planning based on cases A and B are provided in Table 5.2. The optimization based on case A considers the objective of maximizing the minimum reliability index β_{min} during a predefined lifetime $t_{life,d}$ of 50 years. The objective of the optimization for case B corresponds to the maximization of the expected reliability index $E(\beta)$ during the predefined lifetime $t_{life,d}$. When the optimum maintenance times t^*_{ma} for cases A and B are applied, the reliability index profiles are shown in Figure 5.3. The optimum maintenance application times $t^*_{ma,A}$ for case A ranges from $t^{*-}_{ma,A}$ = 25 years to $t^{*+}_{ma,A}$ = 30 years which are associated with the minimum reliability index $\beta_{min,A}$ of 3.5 during the lifetime $t_{life,d}$ [see Figure 5.3(a)]. The maintenance application time before $t^{*-}_{ma,A}$ or after $t^{*+}_{ma,A}$ produces a lower minimum reliability index than 3.5. As shown in Figure 5.3(b), the optimum maintenance time $t^*_{ma,B}$ for case B is equal to 25 years, which results in the maximum expected reliability index $E(\beta) = 5.0$ during the predefined lifetime $t_{life,d}$. Table 5.2 presents the comparisons among the optimum maintenance application time, minimum reliability index and expected reliability index for cases A and B. From Figure 5.3 and Table 5.2, it can be found that (a) the optimum maintenance application time for case A is at least that of case B (i.e., $t^*_{ma,A} \geq t^*_{ma,B}$), (b) the maximum expected reliability index for case A is at most that of case B [i.e., $E(\beta_A) \leq E(\beta_B)$], and (c) the minimum reliability indices for cases A and B are the same (i.e., $\beta_{min,A} = \beta_{min,B} = 3.5$).

Table 5.2   Formulation and results of optimization problem for cases A and B

		Case A	Case B
Formulation of optimization problem	Objective	Maximizing β_{min} during $t_{life,d}$ of 50 years	Maximizing $E(\beta)$ during $t_{life,d}$ of 50 years
	Design variable	Maintenance application time $t_{ma,A}$	Maintenance application time $t_{ma,B}$
	Constraints	$\beta_{th} \leq \beta(t) \leq \beta_{in}$; and $\beta_{th} \leq \beta(t') \leq \beta_{in}$	
	Given conditions	β_{in} = 6.0; β_{th} = 2.0; $\Delta\beta$ = 2; r_β = 0.1/year; t_{ini} = 5 years	
Results of optimization problem	Optimum maintenance application time t^*_{ma}	25 years $\leq t^*_{ma,A}$ $\leq$ 30 years	$t^*_{ma,B}$ = 25 years
	Minimum reliability index β_{min}	$\beta_{min,A}$ = 3.5	$\beta_{min,B}$ = 3.5
	Expected reliability index $E(\beta)$ during $t_{life,d}$	$4.8 \leq E(\beta_A) \leq 5.0$	$E(\beta_B)$ = 5.0

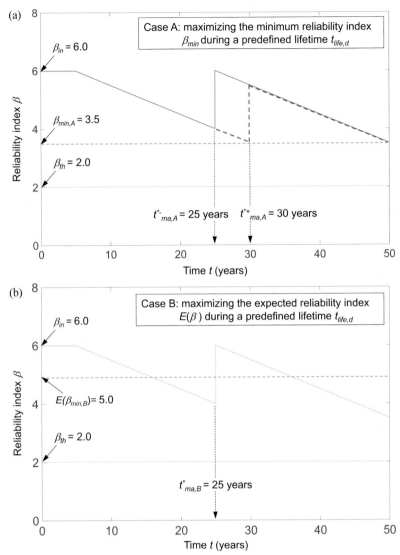

Figure 5.3 Reliability index profiles for optimum maintenance plans based on: (a) case A (maximizing the minimum reliability index β_{min} during a predefined lifetime $t_{life,d}$ of 50 years); and (b) case B (maximizing the expected reliability index $E(\beta)$ during a predefined lifetime $t_{life,d}$ of 50 years).

In general, an increase in the bridge performance during a predefined lifetime requires an increase in both the initial and maintenance costs. For this reason, the performance-based objectives are related with the cost-based objectives considering the initial and maintenance costs. Therefore, both the performance-based objectives and the cost-based objectives have to be considered in the formulation of a multi-objective optimization for bridge management (Okasha and Frangopol 2009b, 2010b).

5.3.2 Cost-Based Objectives

The total life-cycle cost C_{life} consisting of the initial cost C_{ini}, inspection cost C_{ins}, maintenance cost C_{ma}, and failure cost C_{fail} [see Eq. (1.1)] is one of the most representative objectives for optimum bridge management planning (Frangopol 2011; Frangopol and Soliman 2016; Biondini and Frangopol 2016; Frangopol and Kim 2019). The general concept of the life-cycle cost is provided in Section 1.2.2. The optimization associated with minimization of C_{life} can be applied for both bridge design and management as mentioned previously. Through the optimization process based on minimizing C_{life}, the types and times of inspection and maintenances can be optimized.

The failure cost C_{fail}, which is the product of the probability of failure $p_{f,life}$ during a predefined lifetime $t_{life,d}$ and the expected monetary loss C_f as indicated in Eq. (1.2), can be used as an objective for optimum bridge management planning (Orcesi et al. 2010; Orcesi and Frangopol 2011c). Comparing with the total life-cycle cost C_{life}, the failure cost C_{fail} does not take into account the initial cost C_{ini}, inspection cost C_{ins}, and maintenance cost C_{ma}. However, the effect of the inspection and maintenance on the probability of failure p_f is considered to compute the failure cost C_{fail}. Therefore, it is possible to use the objective of minimizing the failure cost C_{fail}. Furthermore, the objective of minimizing the failure cost C_{fail} can be used together with the objective of minimizing inspection cost C_{ins} and/or maintenance cost C_{ma} in the formulation of the multi-objective optimization (Orcesi and Frangopol 2011c; Zhu and Frangopol 2013a; Kim et al. 2013; Soliman et al. 2016).

The cost-based objectives are affected by the discount rate of money, with which the value of the future cost is discounted to the present value (NCHRP 2003). The total life-cycle cost C_{life} during a predefined lifetime $t_{life,d}$ of Eq. (1.1) is modified to compute its present value as (Frangopol et al. 1997b; Enright and Frangopol 1999b; Estes and Frangopol 2005)

$$C_{life} = C_{ini} + \sum_{i=1}^{N_{ins}} \left[\frac{C_{ins,i}}{\left(1+r_{dis}\right)^{t_{ins,i}}} \right] + \sum_{i=1}^{N_{ma}} \left[\frac{C_{ma,i}}{\left(1+r_{dis}\right)^{t_{ma,i}}} \right] + \frac{C_{fail}}{\left(1+r_{dis}\right)^{t_{life,d}}} \qquad (5.6)$$

where r_{dis} = annual discount rate of money. An increase in the discount rate r_{dis} results in reduction of the total life-cycle cost C_{life}. The annual discount rate r_{dis} can affect the optimum maintenance management planning (Enright and Frangopol 1999b; Val and Stewart 2003; Neves et al. 2004; Yang et al. 2006a). For example, three cases (i.e., cases 1, 2 and 3) are considered, which are associated with a single maintenance applied at time $t_{ma,1}$, $t_{ma,2}$ and $t_{ma,3}$, as shown in Figure 5.4(a). It is assumed that (a) the inspection and maintenance are applied at the same time (i.e., $t_{ins,i} = t_{ma,i}$), (b) each case results in the same improvement of the reliability index after maintenance, and (c) the costs C_{ini}, $C_{ins,i}$, $C_{ma,i}$ and C_{fail} during the lifetime $t_{life,d}$ are the same for the three cases considered. Therefore, as shown in Figure 5.4(b), the total life-cycle costs for the three cases are the same ($C_{life,1} = C_{life,2} = C_{life,3}$ where $C_{life,i}$ is the total life-cycle cost of case i) when the annual discount rate of money r_{dis} is ignored (i.e., $r_{dis} = 0$). However, if r_{dis}

is taken into account (i.e., $r_{dis} > 0$), case 3 is the most cost-effective among the three cases, since this case is associated with the largest maintenance application time (i.e., $t_{ma,1} < t_{ma,2} < t_{ma,3}$ and $C_{life,1} > C_{life,2} > C_{life,3}$).

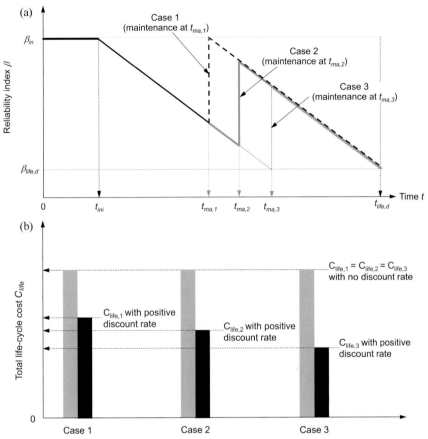

Figure 5.4   Reliability profiles and life-cycle cost associated with cases 1, 2 and 3: (a) reliability profiles; and (b) total life-cycle cost.

The annual discount rate of money generally ranges from 0 to 7% (Sánchez-Silva et al. 2016). The annual discount rates should be determined by considering a number of economic, social and political factors (Val and Stewart 2003). Rackwitz (2006) and Streicher et al. (2008) provide extensive descriptions on the determination of the annual discount rate of money.

5.3.3   Damage Detection-Based Objectives

Instantaneous damage detection can lead to timely and appropriate maintenance, and cost-effective and efficient service life management for deteriorating bridges under uncertainty (Kim and Frangopol 2011c, 2011d; Kim et al. 2013). Therefore, optimum bridge management planning should consider the objectives based on the concept of the probabilistic damage detection (Frangopol and Kim 2019). Damage detection-based objectives for optimum bridge management planning include maximizing the probability of damage detection (see Eq. (3.17)), minimizing the damage detection time and delay (see Eqs. (3.22) and (3.24)), and minimizing the damage detection-based probability of failure (see Eq. (3.27)). The formulations of these objectives are addressed in Section 3.5.

The maintenance delay is defined as the time interval between damage initiation and maintenance application. Based on the event tree used to formulate the expected total inspection and maintenance cost and the expected extended service life (see Figure 4.15), the maintenance delay t_{mdl} can be expressed as (Kim et al. 2011, 2013)

$$t_{mdl} = t_{life,0} - t_{ini} \qquad \text{for} \quad t_{insp} > t_{life,0} \qquad (5.7a)$$

$$t_{mdl} = (1 - P_{det}) \cdot (t_{life,0} - t_{ini}) + P_{det} \cdot t'_{mdl} \quad \text{for} \quad t_{insp} \leq t_{life,0} \qquad (5.7b)$$

where $t_{life,0}$ is the initial service life, t_{ini} is the damage initiation time, P_{det} is the probability of damage detection associated with an inspection method applied at time t_{insp}, and t'_{mdl} is the maintenance delay after damage detection by the inspection at time t_{insp}. Eq. (5.7a) is associated with Branch 1 of the event tree (see Figure 4.15), which represents the event that there will be no inspection and maintenance within the service life (i.e., $t_{insp,1} \geq t_{life,0}$). Two events when the inspection is performed within the service life (i.e., $t_{insp,1} < t_{life,0}$) are integrated in Eq. (5.7b). The first event is associated with no damage detection and no maintenance, which is represented by Branch 2 in Figure 4.15. The second event considers the damage detection and maintenance application, which are represented by Branches 3, 4 and 5. The maintenance delay after damage detection t'_{mdl} is expressed as (Estes and Frangopol 2001; Kim et al. 2011)

$$t'_{mdl} = (1 - P_{ma}) \cdot (t_{life,0} - t_{ini}) + P_{ma} \cdot (t_{insp} - t_{ini}) \qquad (5.8)$$

where P_{ma} is the probability of applying maintenance. As shown in Figure 4.16, several approaches (e.g., delayed, proactive, linear and idealized) to represent P_{ma} can be applied, where P_{ma} depends on the degree of detected damage. Considering the uncertainties associated with the initial service life $t_{life,0}$ and damage initiation time t_{ini}, the maintenance delay t'_{mdl} in Eq. (5.8) should be treated as a random variable.

The objective of minimizing the expected maintenance delay $E(t_{mdl})$ can be used for optimum bridge management. This optimization process provides the optimum types and times of inspections and maintenances (Kim et al. 2011, 2013; Frangopol and Kim 2019). For given damage initiation time t_{ini} and service life $t_{life,0}$, a reduction of the expected maintenance delay $E(t_{mdl})$ can be achieved through an increase in the probability of damage detection of an inspection method P_{det} and/or an increase in the probability of applying maintenance P_{ma}. This reduction is related to an increase in cost. Therefore, in order to compromise these conflicting objectives, multi-objective optimization can be adopted (Kim et al. 2011, 2013).

5.3.4 Service Life-Based Objectives

The optimum bridge management planning can be based on the objectives of maximizing the expected extended service life $E(t_{life,ex})$ (Frangopol and Kim 2019). The formulation of the extended service life uses the event tree of Figure 2.14, which considers the uncertainties associated with the damage propagation prediction, damage detection and degree of detected damage as presented in Section 2.4.1. The optimization with the objective of maximizing $E(t_{life,ex})$ can provide the types and times of inspections and maintenances. An increase in the expected extended service life of a deteriorating bridge can be caused by an increase in the probability of damage detection associated with an inspection method and/or an increase in the number of inspections and maintenance interventions (Kim et al. 2011; Barone and Frangopol 2013).

When the formulations of the objectives consider the relations among damage propagation, damage detection, and maintenance applications under uncertainty, the associated optimization generally requires a large computational time. In order to improve the computational efficiency during the optimization, the lifetime functions including survival function, hazard function, cumulative hazard function, and mean residual life function can be used. The concepts and theoretical background of the lifetime functions are described in Section 1.3.3. The lifetime functions with maintenance effects are investigated in Section 4.3.2. During recent decades, the lifetime functions have been used for probabilistic optimum bridge maintenance management. For example, Yang et al. (2006a, 2006b) determined the most cost-effective maintenance strategies, among possible maintenance scenarios, where the failure cost is computed by using the Weibull lifetime function. The multi-objective optimization integrating availability, redundancy and life-cycle cost, which are based on Weibull lifetime function, was investigated by Okasha and Frangopol (2010b).

5.3.5 Risk-Based Objectives

The risk is generally expressed as a product of the occurrence probability of an adverse event and the resulting monetary loss (Zhu and Frangopol 2012; Dong and Frangopol 2016a). The occurrence probability of the adverse event can be represented by the probability of failure of a component, probability of system failure due to the component failure, and probability of exceeding a damage state (Zhu and Frangopol 2012; Dong et al. 2014b). The monetary loss due to the occurrence of an adverse event can be divided into direct loss and indirect loss (Zhu and Frangopol 2012; Barone and Frangopol 2014a). The direct monetary loss is related to the cost required to replace the failed (or damaged) component and system. The indirect monetary loss includes safety loss, commercial loss and environmental loss. The safety loss quantifies damage inflicted to persons (e.g., fatalities, injuries or general health issues) caused by the system failure. Environmental loss indicates contamination of the environment after the system failure (Hessami 1999). Figure 5.5 shows the event tree associated with the direct and indirect risk caused by the failure of component i resulting in two complimentary events (i.e., system failure and system survival). Therefore, the total risk due to the failure of component i at time t is

$$R_i(t) = p_{f,i}(t) \cdot \left[1 - p_{f,sub|i}(t)\right] \cdot C_{dir,i}(t) + p_{f,i}(t) \cdot p_{f,sub|i}(t) \cdot \left[C_{dir,i}(t) + C_{ind,i}(t)\right]$$

$$= p_{f,i}(t) \cdot C_{dir,i}(t) + p_{f,i}(t) \cdot p_{f,sub|i}(t) \cdot C_{ind,i}(t) \qquad (5.9)$$

$$= R_{dir,i}(t) + R_{ind,i}(t)$$

where $p_{f,i}$ is the probability of failure of the ith component, $p_{f,sub|i}$ is the probability of system failure due to the failure of the ith component, and $C_{dir,i}$ and $C_{ind,i}$ are the direct and indirect monetary losses caused by the failure of component i, respectively. The direct risk $R_{dir,i}(t)$ and indirect risk $R_{ind,i}(t)$ in Eq. (5.9) are expressed, respectively, as (Zhu and Frangopol 2012, 2013a)

$$R_{dir,i}(t) = p_{f,i}(t) \cdot C_{dir,i}(t) \qquad (5.10a)$$

$$R_{ind,i}(t) = p_{f,i}(t) \cdot p_{f,sub|i}(t) \cdot C_{ind,i}(t) \qquad (5.10b)$$

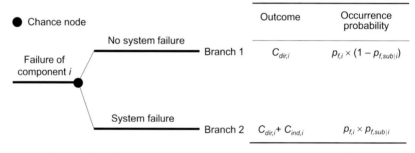

Figure 5.5   Event tree for risk model due to a component failure.

The difference between the probability of failure and risk depends on the monetary loss resulting from the failure of a component as presented in Section 1.4 (see Figure 1.13). Zhu and Frangopol (2012) investigates the effects of (a) the deterioration process of components, (b) the type of system modeling (i.e., series, parallel and series-parallel system), and (c) the correlations among the failure modes of components on the time-dependent direct, indirect and total risk of a bridge system. The effects of maintenance interventions on the risk profile of a deteriorating bridge are illustrated in Figure 1.14. Based on Eq. (5.9) and (5.10), the risk-based robustness index RB_{risk} can be formulated as indicated in Eq. (1.49). Suitable objectives for optimum bridge management include the minimization of total risk and the maximization of risk-based robustness (Saydam et al. 2013a).

5.4   OPTIMIZATION AND DECISION MAKING FOR LIFE-CYCLE BRIDGE MANAGEMENT PLANNING

The life-cycle bridge management can be optimized using the objectives presented in Section 5.3 individually or together (Frangopol and Kim 2019). The general formulation of the optimization problem consisting of multiple objectives is provided in Eq. (5.3). Depending on the objectives, the Pareto optimal solutions of the bi-objective optimization are generally distributed as shown in Figure 5.6. It has been known that the multi-objective optimization can provide multiple well-balanced solutions. Finally, the bridge managers can have flexibility in determining the best Pareto optimal solution.

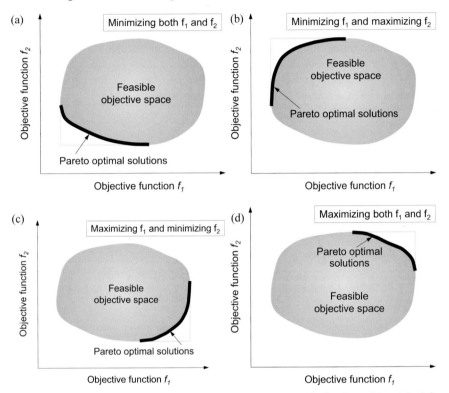

Figure 5.6 Pareto optimal solutions of bi-objective optimization: (a) minimizing both objectives f_1 and f_2; (b) minimizing f_1 and maximizing f_2; (c) maximizing f_1 and minimizing f_2; (d) maximizing both objectives f_1 and f_2.

5.4.1 Correlation among Objectives

The objectives for optimum bridge management may be correlated, since the formulations of the objectives are based on the probabilistic performances and service life prediction, and cost estimation with maintenance interventions. The linear relationship between the two objective functions f_i and f_j can be represented by the Pearson's correlation coefficient ρ_p (f_i, f_j), which is computed as (Ang and Tang 2007)

$$\rho_p\left(f_i, f_j\right) = \frac{E\left[\left\{f_i\left(\mathbf{X}\right) - E\left[f_i\left(\mathbf{X}\right)\right]\right\} \cdot \left\{f_j\left(\mathbf{X}\right) - E\left[f_j\left(\mathbf{X}\right)\right]\right\}\right]}{\sigma_i \cdot \sigma_j} \tag{5.11}$$

where $\mathbf{X}$ is the vector of design variables, E is the expected value, and σ_i and σ_j are the standard deviations (SDs) of the objective values for f_i and f_j in the design space, respectively. The monotone relation between f_i and f_j is quantified using the Spearman's rank correlation coefficient $\rho_s(f_i, f_j)$ as (Myers et al. 2003)

$$\rho_s(f_i, f_j) = 1 - \frac{6\sum_{i=1}^{n} d_i^2}{n(n^2 - 1)} \qquad (5.12)$$

where n = number of observations (i.e., number of values of objective functions f_i (or f_j) in the design space); and d_i = difference between the ranks of each observation (i.e., the values of the objective functions f_i and f_j).

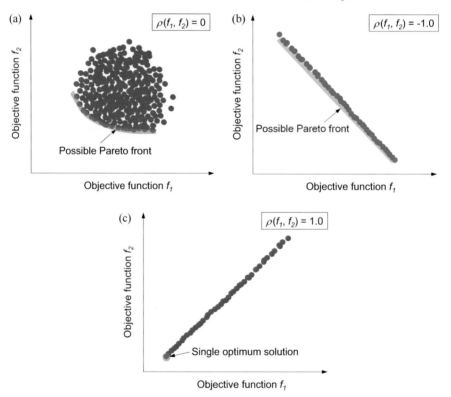

Figure 5.7 Possible Pareto front and correlation of minimizing both objectives f_1 and f_2: (a) $\rho(f_1, f_2) = 0.0$; (b) $\rho(f_1, f_2) = -1.0$; and (c) $\rho(f_1, f_2) = 1.0$.

Figure 5.7 shows the possible Pareto fronts when the coefficients of correlation $\rho(f_1, f_2)$ defined in Eqs. (5.11) and (5.12) are 0.0, −1.0, and 1.0. Figure 5.7(a) shows the possible Pareto front when the objectives of minimizing f_i and f_j are independent. When the coefficient of correlation $\rho(f_1, f_2)$ is equal to −1.0, the objectives of minimizing f_1 and f_2 conflict with each other. The associated Pareto front is shown in Figure 5.7(b). The coefficient of correlation $\rho(f_1, f_2)$ equal to 1.0

indicates that the objective of minimizing f_1 supports the objective of minimizing f_2, and a single optimum solution is obtained instead of the Pareto front (see Figure 5.7(c)). In this case, the single-objective optimization based on f_1 (or f_2) is formulated, and a bi-objective optimization is not necessary. Using the correlation coefficient, efficiency in finding the Pareto optimal solutions and decision making can be improved (Kim and Frangopol 2017).

5.4.2   Essential Objectives

The optimization problem consisting of more than three probabilistic objectives requires a high computational cost to find the entire Pareto optimal solutions (Deb and Saxena 2006; Saxena et al. 2013). The essential objectives among the initial objectives affect the Pareto optimal solutions, while the redundant objectives do not affect the Pareto optimal solutions. Figure 5.8 compares the Pareto optimal solutions associated with the initial objectives (i.e., minimizing $f_1, f_2, \ldots,$ and f_n) and essential objectives (i.e., minimizing $f_1, f_2,$ and f_3). From this figure, it can be seen that when the essential objectives are used instead of the initial objectives there will be no change in the Pareto optimal solutions. The approach to identify the essential and redundant objectives among the initial objective set is required to compute the Pareto solutions efficiently. Brockhoff and Zitzler (2006, 2007, 2009) developed such an approach based on the dominance relationships among the objective values.

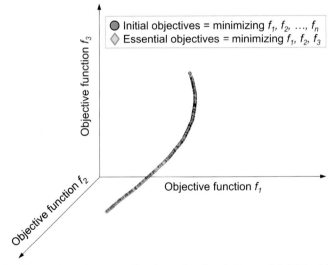

Figure 5.8   Comparison between Pareto optimal solutions of initial objectives and essential objectives.

5.4.3   Multi-Objective Optimization Methods

The multi-objective optimization problem can be solved using the approaches such as multi-objective genetic algorithm, weighted sum method, weighted min-max method, weighted global criterion method, lexicographic method, and bounded objective function method (Arora 2017). The multi-objective genetic algorithm can find all Pareto optimal solutions without requiring continuity or differentiability of the objective functions. For this reason, the genetic algorithm has been applied to solve the multi-objective optimization for bridge performance and service life management, which generally consists of probabilistic objectives with complex conditions (Frangopol 2011). However, the multi-objective genetic algorithm for bridge performance and service life management requires a larger computational cost, as the number of objectives increases and more accuracy is required. Therefore, the genetic algorithm for multi-objective optimization continues to improve its computational efficiency and accuracy (Singh et al. 2011; Saxena et al. 2013).

The weighted sum, weighted min-max, and weighted global criterion methods combine all the objective functions to be minimized to form a single objective function. The objective functions of these methods are expressed as (Arora 2017)

$$f_{ws}(\mathbf{X}) = \sum_{i=1}^{n} w_i f_i^{norm}(\mathbf{X}) \tag{5.13a}$$

$$f_{wmm}(\mathbf{X}) = \max_{i=1}^{n} \left[w_i \left\{ f_i^{norm}(\mathbf{X}) - f_i^{up} \right\} \right] \tag{5.13b}$$

$$f_{wgc}(\mathbf{X}) = \left(\sum_{i=1}^{n} \left[w_i \left\{ f_i^{norm}(\mathbf{X}) - f_i^{up} \right\} \right]^{p} \right)^{1/p} \tag{5.13c}$$

where $f_{ws}(\mathbf{X})$, $f_{wmm}(\mathbf{X})$ and $f_{wgc}(\mathbf{X})$ are the combined objective functions associated with the weighted sum, weighted min-max, and weighted global criterion methods, respectively, n is the number of objectives to be minimized, $\mathbf{X}$ is the vector of design variables, $f_i^{norm}(\mathbf{X})$ is the normalized objective function, w_i is the weight factor, and p is a constant. The utopia point of $f_i^{norm}(\mathbf{X})$ is denoted as f_i^{up}. These methods are computational efficient, but cannot find all the Pareto optimal solutions. More details on the lexicographic method, bounded objective function method, and other multi-objective optimization algorithms and techniques can be found in Coello Coello (2003), Rao (2009), Deb (2001), Marler and Arora (2010), Arora (2017), Cui et al. (2017), and Gunantara (2018).

5.4.4   Multi-Attribute Decision Making

The multi-attribute decision making (MADM) to select the best solution from the Pareto optimal front requires determining the weight of the essential objectives, and estimating the overall assessment value of the Pareto optimal solutions (Kim and Frangopol 2018a, 2018b). The Pareto optimal solutions computed from the multi-objective optimization are used to compute the weights of the essential objectives. The weights of the redundant objectives are predefined as zero, since the redundant objectives do not affect the Pareto optimal solutions. Using the weight of the essential objectives, the overall assessment values of the Pareto optimal solutions are obtained. Finally, the best solution corresponds to the Pareto optimal solution with the largest overall assessment value.

The SD method is one of the most representative weight determination methods. This method is based on the premise that if the objective values of the Pareto optimal solutions have a larger SD, the associated objective needs a larger weight of the objective. Accordingly, the weight of the ith essential objective w_i is estimated as (Deng et al. 2000)

$$w_i = \frac{\sigma_i}{\sum_{j=1}^{n_e} \sigma_j} \tag{5.14}$$

where σ_j = SD of the jth objective values of the Pareto optimal solutions; and n_e = number of essential objectives. The criteria importance through inter-criteria correlation (CRITIC) method, which is another representative weight determination method, assumes that the objective less correlated with other objectives has more impact on the decision making, and more value of the weight. The CRITIC method determines the weight w_i of the ith essential objective as (Diakoulaki et al. 1995)

$$w_i = \frac{\sigma_i \cdot \sum_{j=1}^{n_e} \left(1 - \rho_{ij}\right)}{\sum_{j=1}^{n_e} \left(\sigma_j \cdot \sum_{k=1}^{n_e} \left(1 - \rho_{jk}\right)\right)} \tag{5.15}$$

where ρ_{ij} = correlation coefficient between the values of the ith essential objective and the jth essential objective of the Pareto optimal solutions.

The most widely used MADM methods to compute the overall assessment value of the Pareto optimal solutions include the simple additive weighting (SAW) method, the technique for order preference by similarity to ideal solution (TOPSIS) method, and the elimination and choice expressing the reality (ELECTRE) method (Zanakis et al. 1998; Yeh 2002; Velasquez and Hester 2013; Kabir et al. 2014). The SAW method estimates the overall assessment value of the ith Pareto optimal solution A_i (Yoon and Hwang 1995)

$$A_i = \sum_{j=1}^{n_e} w_j f_{ji}^{norm} \tag{5.16}$$

where f_{ji}^{norm} = jth normalized objective value of the ith Pareto optimal solution. The basic concept of the TOPSIS method is that the best optimal solution has the shortest distance from the positive ideal solution, and the longest distance from the negative ideal solution. Therefore, the overall assessment value of the ith Pareto optimal solution A_i is estimated as (Yoon and Hwang 1981; Hwang et al. 1993)

$$A_i = \frac{d_i^-}{d_i^+ + d_i^-} \tag{5.17}$$

where d_i^+ is the distance between the ith Pareto optimal solution and the positive ideal solution, and d_i^- is the distance between the ith Pareto optimal solution and the negative ideal solution. The detailed algorithm of the TOPSIS method is available in Yoon and Hwang (1981) and Hwang et al. (1993). The ELECTRE method was introduced by Benayoun et al. (1966), and was developed continuously as ELECTRE I, II, III, IV and IS methods (Fei et al. 2019). The ELECTRE method finds the binary outranking relations among the Pareto optimal solutions. Finally, the overall ranking of the Pareto optimal solutions can be estimated. The detailed algorithms and applications of the ELECTRE method are available in Roy (1971), Nijkamp and van Delft (1977), Voogd (1983), Kafandaris (2002), Hatami-Marbini and Tavana (2011), and Govindan and Jepsen (2016).

5.5 CONCLUDING REMARKS

This chapter deals with general concepts and theoretical background for optimum life-cycle bridge inspection and maintenance planning under uncertainty. The optimum bridge design, inspection and maintenance planning can be achieved based on the probabilistic life-cycle performance and cost analysis. The objectives for this probabilistic optimization are categorized into (a) performance-based; (b) cost-based; (c) damage detection-based; (d) service life-based; and (e) risk-based. The multiple objectives can be integrated systematically into multi-objective optimization. In order to find the Pareto optimal solutions of the multi-objective optimization efficiently, the correlation among objectives, identification of essential and redundant objectives, and multi-objective optimization are described. The approaches for multi-attribute decision making, which are used to select the best Pareto optimal solution, are presented. The extensive applications for optimum life-cycle bridge inspection and maintenance planning are provided in Chapter 6.

Chapter 6

Applications of Optimum Life-Cycle Bridge Management Planning

ABSTRACT

Chapter 6 presents applications for optimum life-cycle bridge inspection and management planning. The applications are categorized into three topics: probabilistic bridge performance and service life prediction, probabilistic optimum bridge inspection and monitoring planning, and probabilistic optimum bridge maintenance planning. The presented applications are selected from previous investigations on optimum life-cycle bridge management planning of an individual bridge. The novelties of the investigations associated with the applications are summarized.

6.1 INTRODUCTION

This chapter deals with applications related to the optimum life-cycle bridge management planning under uncertainty. The applications are categorized according to three topics: (I) probabilistic bridge performance and service life prediction, (II) probabilistic optimum bridge inspection and monitoring planning, and (III) probabilistic optimum bridge maintenance planning at an individual bridge-level. The applications associated with topic I show that the probabilistic bridge performance and service life can be predicted using both state functions and lifetime distributions. The applications associated with topic II show that single- and multi-objective optimum bridge inspection and monitoring planning can be formulated considering different objectives. Finally, applications for topic III illustrate that the optimum bridge maintenance planning can be formulated based on multiple objectives that be considered simultaneously for more rational and flexible optimization. In all three topics, it is shown that information from inspection and/or monitoring can be used to update the results.

6.2 PROBABILISTIC BRIDGE PERFORMANCE AND SERVICE LIFE PREDICTION

The probabilistic bridge performance and service life prediction are essential for life-cycle management planning (Frangopol and Soliman 2016; Frangopol et al. 2017; Frangopol and Kim 2019). As described in Section 2.3, the bridge performance prediction can be computed by using state functions and lifetime distributions. The appropriate use of information from inspection and structural health monitoring (SHM) can lead to accurate and reliable bridge performance prediction, as explained in Section 3.3. Several applications are provided in this section.

6.2.1 Bridge Performance Prediction Using State Functions

The time-dependent state function is formulated considering the loading effect and the resistance deterioration based on fatigue and corrosion damage propagations over time. In Frangopol et al. (1997a), the time-dependent bending and shear reliability indices of deteriorating RC T-girders are predicted by considering uniform corrosion in reinforcement bars. The bending and shear reliability index profiles for various thickness losses and corrosion rates are provided. The limit state functions for moment and shear capacities are formulated based on AASHTO (1992) and Lin and Frangopol (1996). In order to compute the reliability index, the Monte Carlo simulation-based program MCREL (Lin 1995) is adopted. This time-dependent reliability index prediction is applied to the optimum design of RC girders based on life-cycle cost (Frangopol et al. 1997b). Stewart and Rosowsky (1998a) developed a probabilistic framework for the time-dependent reliability prediction of RC bridge decks, where the relationships among cracking, diffusion of chlorides, and corrosion initiation/propagation caused by atmospheric exposure in a marine environment are addressed.

The cumulative-time failure probability for an existing RC T-beam bridge (i.e., Colorado Highway Bridge L-18-BG) subjected to corrosion was estimated by Enright and Frangopol (1999c). The time-dependent resistance of a bridge component was computed using the resistance degradation function defined in Mori and Ellingwood (1993) and Enright and Frangopol (1998a). The effects of various parameters related to the models for load and resistance were studied in Enright and Frangopol (1998b). The live load effect based on AASHTO (1994) was implemented to compute the cumulative-time failure probability. The cumulative-time failure probabilities of the series, parallel and series-parallel modelings for a bridge system were compared. Figure 6.1(a) shows the cross-section of Colorado Bridge L-18-BG associated with the illustrative example provided in Enright and Frangopol (1999c). The fault trees in Figures 6.1(b), 6.1(c), and 6.1(d) are used to define system failures. For system I of Figure 6.1(b), the system failure is defined as failure of any girder in a single span.

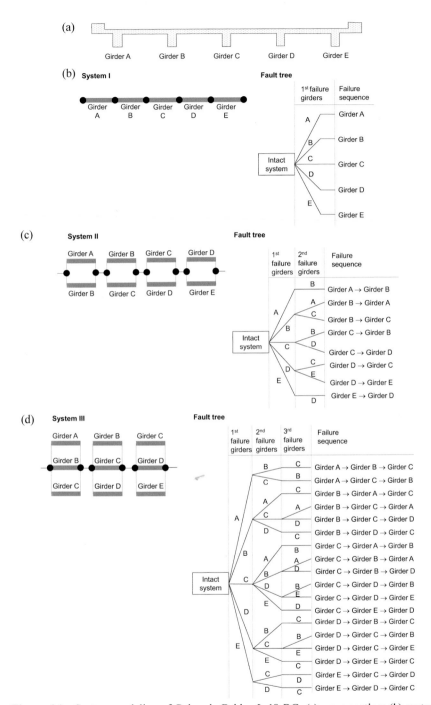

Figure 6.1 System modeling of Colorado Bridge L-18-BG: (a) cross section; (b) system I and associated fault tree; (c) system II and associated fault tree; and (d) system III and associated fault tree (adapted from Enright and Frangopol 1999c).

The series-parallel models for systems II and III (see Figures 6.1(c) and 6.1(d)) represent the system failures which are caused by failure of any two and any three adjacent girders, respectively. These system modelings based on the fault trees take into account the redistribution of live load from failed girders as investigated in Enright and Frangopol (1998c). The Monte Carlo simulation-based program RELTSYS (Enright and Frangopol 2000) was used to estimate the cumulative-time failure probability of the bridge system.

The reliability index predictions for various types of existing bridges (e.g., RC bridges, prestressed concrete bridges, steel girder bridges) were extensively investigated by Akgül and Frangopol (2004b, 2004c, 2005a, 2005b). In Akgül and Frangopol (2004c), the state functions for prestressed concrete bridges are defined, where the critical positive and negative moments, and tensile stress in concrete are considered. For time-dependent reliability analysis, the corrosion penetration in the prestressed concrete girder is predicted. The reliability index prediction in terms of slab flexure, girder flexure and shear, and serviceability of steel bridges subjected to corrosion is presented in Akgül and Frangopol (2004b). Akgül and Frangopol (2005a) derived the limit state equations and presented the corrosion deterioration model for RC bridge slabs and girders. The applications of reliability prediction for existing RC bridges (i.e., Colorado Highway Bridges E-17-HS, E-17-HR and E-17-HE) are provided in Akgül and Frangopol (2005b). The bridge system reliability index in Akgül and Frangopol (2004b, 2004c, 2005a, 2005b) is computed based on the series-parallel system, in which the failure of any two adjacent concrete girders in flexure or shear is treated as the system failure. The state functions are formulated by adopting the requirements and formulas associated with the load and strength estimation based on AASHTO (1996) specification. Time-dependent reliability index profiles for the bridges were obtained using the computer program RELNET (Akgül and Frangopol 2004d).

Darmawan and Stewart (2007) performed accelerated pitting corrosion tests and developed the probabilistic pitting corrosion model for strands of prestressing concrete bridges. The probabilistic corrosion propagation model is based on the outcomes presented in Stewart and Rosowsky (1998a, 1998b) and Vu and Stewart (2000). By integrating the nonlinear finite element analysis with the developed probabilistic models of corrosion initiation and propagation, the time-dependent reliability of the prestressed concrete bridge girder was assessed.

Moreover, the time-variant reliability and redundancy of bridge systems were studied by Okasha and Frangopol (2009a). The effects of the material behavior (i.e., ductile and brittle), resistance deterioration, and load amplification on the time-dependent system reliability and redundancy were also investigated. Zhu and Frangopol (2012) investigated the reliability, redundancy, and risk of an existing steel girder and RC deck bridge (i.e., Colorado Highway Bridge E-17-AH), where the effects of (a) bridge component deterioration, (b) system modeling, and (c) correlations among the failure modes of bridge components on the time-dependent reliability, redundancy, and risk of bridge system were quantified. The cumulative-time failure probabilities and time-dependent risks of girders under traffic loads and pier columns under scour were assessed in Zhu and Frangopol (2016). The approach for the time-dependent reliability analysis taking the hazard associated with airborne chlorides into consideration was presented by Akiyama et al. (2010).

6.2.2 Bridge Performance and Service Life Prediction Using Lifetime Distributions

The lifetime distributions are formulated with an appropriate distribution type (e.g., exponential and Weibull), as described in Section 1.3.3. Yang et al. (2004) applied the lifetime distributions to assess the cumulative-time probability of failure of an existing steel girder and RC deck bridge (i.e., Colorado Highway Bridge E-17-AH), where four failure modes were considered for the bridge system. These four failure modes indicate that the failure of the bridge system is caused by (a) failure of any girder or deck failure, (b) failure of any external girder or failure of any two adjacent internal girders or deck failure, (c) failure of any two adjacent girders or deck failure, and (d) failure of any three adjacent girders or deck failure. The effects of assuming perfectly correlated and statistically independent bridge components and maintenance actions on the cumulative-time probability of system failure were investigated.

Okasha and Frangopol (2010a) presented the application of the lifetime performance prediction of an existing steel girder and RC deck bridge (i.e., Colorado Highway Bridge E-17-AH) based on lifetime distributions. The effects of essential maintenance actions on the lifetime reliability, redundancy, unavailability, failure rate, and cumulative failure rate were investigated. The combined effects of multiple types of maintenance actions (i.e., essential maintenance (EM), and preventive maintenance (PM)) on the unavailability and redundancy are provided in Okasha and Frangopol (2010b). Barone and Frangopol (2014a) predicted the system reliability, availability and hazard functions with essential maintenance actions. The comparisons among the bridge performances based on limit states (i.e., annual reliability and annual risk) and bridge performances based on lifetime distribution (i.e., availability and hazard function) were investigated when the failure modes of components are statistically independent and perfectly correlated.

6.2.3 Bridge Performance and Service Life Prediction Based on Inspection and Monitoring

During the last decades, the probabilistic approaches to use the information from inspection and monitoring for predicting and updating the bridge performance and service life have been developed (Frangopol and Soliman 2016). Enright and Frangopol (1999a) proposed a probabilistic approach to incorporate the existing information associated with prediction of the RC bridge with new inspection data, where the corrosion rate is updated through the Bayesian techniques. Based on this approach the cumulative-time failure probabilities of the bridge system are updated. Estes and Frangopol (2003) examined to what degree the information from the bridge management systems (e.g., PONTIS (Cambridge Systematics, Inc. 2009)) can be used to update the reliability of existing highway bridges subjected to corrosion.

The integration of monitoring data into the structural reliability assessment and prediction and the use of monitored data for the development of structural performance prediction models were introduced in Frangopol et al. (2008a). The long-term monitoring data of an existing bridge in Pennsylvania (i.e., Lehigh River Bridge SR-33) were used. Figure 6.2 shows the reliability index for the yield strength of structural components of the Lehigh River Bridge SR-33, which were obtained from the monitoring data during the construction period. The reliability index based on monitoring data is assessed and predicted in terms of the yield strength and fatigue resistance of the steel truss members. A general approach to develop the performance prediction functions based on monitored extreme data and to assess the associated reliability index was proposed by Frangopol et al. (2008b). Furthermore, an effective approach to update the performance prediction functions and reliability using the Bayesian theorem with monitoring data was introduced by Strauss et al. (2008).

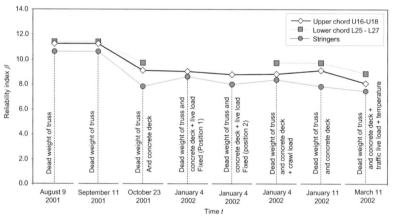

Figure 6.2   Reliability index for the yield strength during construction period of the Lehigh River Bridge SR-33 (adapted from Frangopol et al. 2008a).

Liu et al. (2009a) proposed an approach to evaluate safety of existing bridges based on monitored structural responses and component conditions. A state function is formulated with the measured strain data from SHM. The new concepts of the condition function and prediction function are introduced. To estimate the stresses at the non-monitored locations, the condition function is utilized, considering the condition assessment outcomes, strain gauge locations, and failure modes under consideration. The prediction function is used to predict the extreme stress induced by truck traffic. Based on this investigation, the safety of an existing bridge over the Wisconsin River (i.e. I-39 Bridge) was assessed by Liu et al. (2009b), where the failure is represented by the probability that the monitored maximum stresses induced by the actual heavy vehicles exceed the predefined target stress limit. The sensitivity studies with respect to the system modeling, coefficients of correlation among random variables, types of probability density functions (PDFs) for the monitored extreme stresses, measurement errors, and number of monitored extreme stresses are reported.

The integration of SHM data into fatigue reliability assessment and prediction was presented by Liu et al. (2010a). The monitoring program was performed to estimate the fatigue performance of the details of a retrofitted steel bridge (i.e., Birmingham Bridge) in Pittsburgh, Pennsylvania. The finite element modeling was used to identify the fatigue critical locations after retrofitting, and to modify the original SHM data when the identified fatigue critical locations are different from the monitored locations. Furthermore, Kwon and Frangopol (2010) estimated the fatigue reliability of existing steel bridges (i.e., Neville Island Bridge and Birmingham Bridge in Pittsburgh), where a linear stress life approach (or a single linear *S-N* approach) was adopted to predict the equivalent stress ranges based on monitoring data. The fatigue life of steel bridges was predicted using monitoring data and a bilinear stress life approach (or a bilinear *S-N* approach) by Kwon et al. (2012) and Soliman et al. (2013b).

The monitoring data can be used to predict the serviceability lifetime performance of existing steel girder bridges based on AASHTO (2007b) as indicated by Orcesi and Frangopol (2010). This approach uses the data from crawl tests and long-term monitoring at sensor locations. The data from the crawl tests update the distribution factors of each girder of the bridge, and the long-term monitoring data provide the extreme values of bending moment induced by trucks. These two types of monitoring data are used to formulate and update the state functions for the serviceability lifetime performance. Based on the study of Orcesi and Frangopol (2010), Orcesi and Frangopol (2013) predicted serviceability and safety, and analyzed the impact of short monitoring interruptions on the accuracy of the performance assessment.

Okasha et al. (2012) showed that the data from the crawl tests can be used to update the finite element model and lifetime reliability, and finally, the lifetime distributions of the bridge components can be generated. Furthermore, the integrated framework for bridge management based on SHM was proposed by Okasha and Frangopol (2012). This proposed framework consists of modeling a bridge for finite element analysis, performing the bridge life-cycle performance analysis, updating the existing bridge life-cycle performance, and optimizing the maintenance strategy. Several modeling tools and techniques such as incremental nonlinear finite element analyses, quadratic response surface modeling using the design of experiments concepts, and Latin hypercube sampling are used in this framework.

6.3   PROBABILISTIC OPTIMUM BRIDGE INSPECTION AND MONITORING PLANNING

The probabilistic objectives for optimum inspection and maintenance planning are provided in Section 5.3. Through the single- and multi-objective optimization (MOOP) process and decision making, the inspection planning, monitoring planning and combined inspection and monitoring planning can be optimized (Frangopol and Kim 2019). Moreover, the updating process after each inspection and monitoring can improve the accuracy and reliability of the inspection and monitoring planning (Kim et al. 2019). This section presents the applications associated with optimum inspection and monitoring planning with updating based on inspection and monitoring information.

6.3.1   Optimum Inspection Planning

The optimum life-cycle inspection planning of bridges under uncertainty based on minimization of the expected damage detection delay, defined as the expected time-lapse since a bridge has been damaged until the damage is detected by inspection, was introduced by Kim and Frangopol (2011c). The corrosion of RC bridges is treated as critical damage. A damage detectability function is used to assess the quality of inspection method according to damage intensity. The single objective optimization problem results in the optimum inspection times for given corrosion damage propagation and inspection method. The proposed formulation is also used for monitoring planning. The approach is applied to an existing steel girder and RC deck bridge (i.e., Colorado Highway Bridge E-17-AH). This study shows that an increase of the number inspection and/or monitoring actions, and improvement of inspection quality reduces the expected damage detection delay, but needs additional cost. In order to find well-balanced solutions of these two conflicting criteria, bi-objective optimization was applied by minimizing both the expected damage detection delay and inspection cost (Kim and Frangopol 2011d). The Pareto solutions of this bi-objective optimization indicate the optimum number and quality of inspections as well as the optimum inspection times. The cost-based optimum scheduling of inspection and monitoring for fatigue-sensitive structures developed in Kim and Frangopol (2011b) is based on the minimization of the damage detection-based probability of failure. It is shown that the failure cost significantly affects the optimum solutions. Soliman et al. (2013a) investigated optimum inspection planning for multiple fatigue sensitive details of steel bridges, where the probability of damage detection for multiple details should be maximized. A probabilistic optimum inspection planning introduced by Soliman et al. (2016) is based on simultaneously minimizing the expected life-cycle cost, maximizing the expected extended service life, and minimizing the expected maintenance delay for fatigue sensitive structures.

The decision-making framework for the multi-objective optimum inspection planning of fatigue-sensitive bridges was proposed by Kim and Frangopol (2018a), where six objectives (i.e., maximizing the probability of damage detection, minimizing the expected damage detection delay, minimizing the expected maintenance delay, minimizing the damage detection-based probability of failure, maximizing the expected extended service life, and minimizing the expected life-cycle cost) were considered. Also, an efficient decision making strategy, including identification of the essential objectives, determination of the weight factor of the essential objectives, and selection of the best Pareto optimal solution (see Section 5.4), was proposed.

Considering these objectives, Frangopol and Kim (2019) solved the bi-, tri-, quad- and six-objective optimization problems associated with a fatigue sensitive

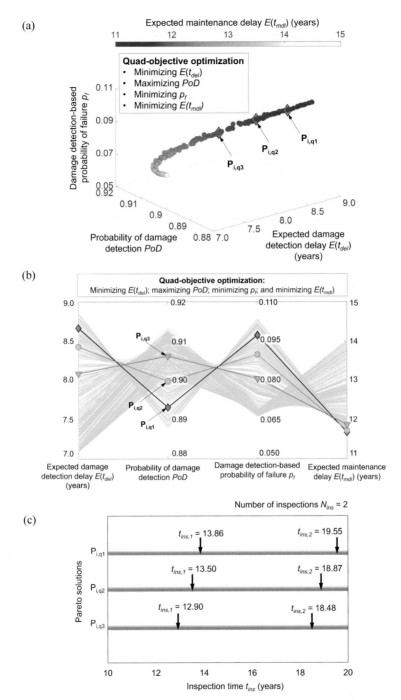

Figure 6.3 Quad-objective optimum inspection planning: (a) Pareto front in 3D Cartesian coordinates; (b) Pareto front in parallel coordinates; and (c) Optimum inspection times of the representative Pareto solutions $P_{i,q1}$, $P_{i,q2}$ and $P_{i,q3}$ in (a) and (b).

detail of a steel bridge (I-64 Kanawha River Bridge) located in West Virginia, and compared the Pareto optimum solutions associated with these problems. The Pareto front for the quad-objective optimum inspection planning is illustrated in the 3D Cartesian coordinate system and the parallel coordinate system as shown in Figures 6.3(a) and 6.3(b), respectively. The quad-objective optimization is based on minimizing the expected damage detection delay $E(t_{del})$, maximizing the probability of damage detection PoD, minimizing the damage detection-based probability of failure p_f, and minimizing the expected maintenance delay $E(t_{mdl})$. One solution in Figure 6.3(a) is represented by the polyline connecting the objective values on the vertical axes in Figure 6.3(b). The optimum inspection times of the representative Pareto solutions $P_{i,q1}$, $P_{i,q2}$ and $P_{i,q3}$ in Figures 6.3(a) and 6.3(b) are illustrated in Figure 6.3(c). The objective values and inspection times of the solutions $P_{i,q1}$, $P_{i,q2}$ and $P_{i,q3}$ are also provided in Table 6.1. For example, the solution $P_{i,q1}$ in Figure 6.3(a) is expressed as the polyline with downward-pointing triangle markers in Figure 6.3(b), where the associated objective values are $E(t_{del})$ = 8.67 years, PoD = 0.89, p_f = 0.097, and $E(t_{mdl})$ = 11.67 years (see also Table 6.1). The solution $P_{i,q1}$ requires two inspections at 13.86 years and 19.55 years as shown in Figure 6.3(c) and Table 6.1. Moreover, the Pareto front of the six-objective optimization is illustrated in the parallel coordinate system as shown in Figure 6.4. The six-objective optimization is formulated by adding two objectives (i.e., maximizing the expected extended service life $E(t_{life})$, and minimizing the expected life-cycle cost $E(C_{life})$) to the quad-objective optimization of Figure 6.3. The objective values and inspection times associated with the representative solutions $P_{i,q4}$, $P_{i,q5}$ and $P_{i,q6}$ in Figure 6.4 are provided in Table 6.1.

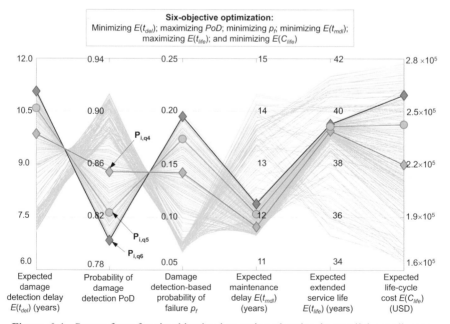

Figure 6.4 Pareto front for six-objective inspection planning in parallel coordinates.

Table 6.1 Objective function values and inspection times associated with Pareto optimum solutions in Figures 6.3 and 6.4.

Pareto optimum solution	Objective function values						Optimum inspection times (years)	
	Probability of damage detection PoD	Expected damage detection delay $E(t_{del})$ (years)	Damage detection–based probability of failure p_f	Expected main-tenance delay $E(t_{mdl})$ (years)	Expected extended service life $E(t_{life})$ (years)	Expected total life-cycle cost $E(C_{life})$ (USD)	$t_{ins,1}$	$t_{ins,2}$
$P_{i,q1}$	0.89	8.67	0.097	11.67	–	–	13.86	19.55
$P_{i,q2}$	0.90	8.43	0.090	11.72	–	–	13.50	18.87
$P_{i,q3}$	0.91	8.09	0.081	11.87	–	–	12.90	18.48
$P_{i,q4}$	0.85	9.86	0.14	11.82	39.27	219849	15.58	19.97
$P_{i,q5}$	0.82	10.58	0.17	12.06	39.11	242604	16.51	19.98
$P_{i,q6}$	0.80	11.06	0.20	12.25	39.52	259257	17.09	19.99

6.3.2 Optimum Monitoring Planning

The bridge monitoring planning can be optimized based on various objectives. The objectives can be formulated with and without information related to damage initiation and propagation prediction (Kim and Frangopol 2020). If the information on damage initiation and propagation is not enough, the bi-objective optimization based on maximizing the expected average availability of monitoring data and minimizing the total monitoring cost can be applied (Kim and Frangopol 2011a). This is because the expected average availability consists of monitoring duration and prediction durations as indicated in Eq. (3.16). The proposed bi-objective optimization was applied to an existing five-span continuous steel plate girder bridge (i.e., Bridge 37–75) in Wisconsin, monitored for a total of approximately 95 days with 24 strain gages and two displacement sensors (Mahmoud et al. 2005). This bi-objective optimization provides the optimum monitoring plan with a uniform time interval. Figure 6.5(a) shows the Pareto front of the bi-objective optimization based on maximizing the expected average availability of monitoring data and minimizing the total monitoring cost. The monitoring plans associated with the solutions $P_{m,A1}$, $P_{m,A2}$, $P_{m,A3}$ and $P_{m,A4}$ in Figure 6.5(a) are illustrated in Figure 6.5(b). An approach for optimum monitoring planning according to the time-dependent reliability importance factor of an individual bridge component was developed by Kim and Frangopol (2010). As a result, the non-uniform time interval of monitoring planning can be achieved. Furthermore, in order to incorporate the effect of the decision maker's risk attitude on the desirability of monitoring planning, the utility theory was adopted by Sabatino and Frangopol (2017), where the two objectives (i.e., maximizing the utilities associated with (a) the expected average availability of the prediction model and (b) monitoring cost) are formulated to find the monitoring duration and prediction duration for both uniform and non-uniform time intervals.

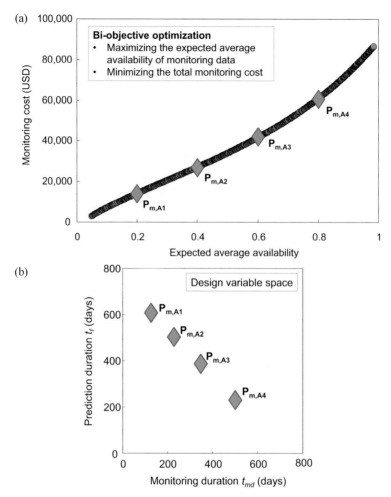

Figure 6.5 Optimum monitoring planning based on maximizing the expected average availability of monitoring data and minimizing the total monitoring cost: (a) Pareto front and four optimal solutions; and (b) optimum monitoring and prediction durations for the four Pareto solutions in (a) (adapted from Kim and Frangopol 2010).

When enough information for damage initiation and propagation is available, the monitoring planning can be based on various objectives such as minimizing the expected damage detection delay, minimizing the expected maintenance delay, maximizing the reliability index, maximizing the expected service life extension and minimizing the expected life-cycle cost (Kim and Frangopol 2018b; 2020). Kim and Frangopol (2018b) presented an efficient approach for integrating multiple objectives (i.e., minimizing the expected damage detection delay, minimizing the expected maintenance delay, maximizing the reliability index, maximizing the expected service life extension, and minimizing the expected life-cycle cost) for optimum monitoring planning. Through the optimization process for monitoring planning, the number of monitorings, monitoring starting times and monitoring durations can be optimized. As an illustrative example, the RC deck of the I-39 Northbound Bridge in Wisconsin is used for the quad-objective optimum monitoring planning as shown in Figure 6.6. Figures 6.6(a) shows the Pareto front of the quad-objective optimization consisting of minimizing the expected damage detection delay, minimizing the expected maintenance delay, maximizing the expected service life extension, and minimizing the expected life-cycle cost. Figure 6.6(b) provides the optimum monitoring plans of the representative Pareto solutions $P_{m,q1}$, $P_{m,q2}$ and $P_{m,q3}$ in Figure 6.6(a), where the optimum monitoring plan indicates the monitoring starting times for given number of monitorings $N_{mon} = 2$ and monitoring duration $t_{md} = 0.5$ years. The generalized computational platform for the multi-objective optimum monitoring planning and decision making was presented in Kim and Frangopol (2020).

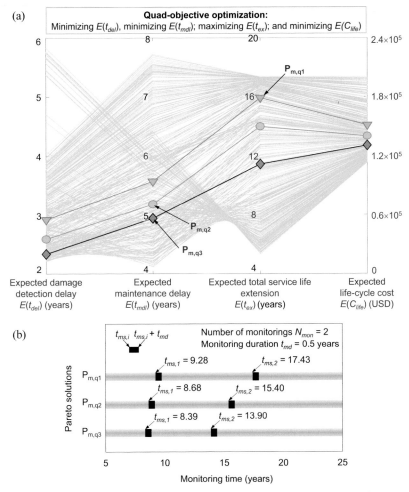

Figure 6.6 Optimum monitoring planning: (a) Pareto optimal solutions of the quad-objective optimization in parallel coordinates; (b) monitoring times and durations of solutions $P_{m,q1}$, $P_{m,q2}$ and $P_{m,q3}$ in (a).

6.3.3   Optimum Combined Inspection and Monitoring Planning

The optimum combined inspection and monitoring planning can be achieved by minimizing both the expected damage detection delay and expected cost for inspection and monitoring (Kim and Frangopol 2012). When monitoring is applied after an inspection, the formulation of damage detection delay is presented in Eq. (3.25). In this manner, the expected damage detection delay associated with more combinations of inspection and monitoring can be formulated. The procedure to find the final Pareto front of the bi-objective optimization when multiple inspection and monitoring actions are considered is provided in Kim and Frangopol (2012). The proposed procedure is applied to an existing truss bridge (i.e., Yellow Mill Pond Bridge) located in Connecticut. The bi-objective optimization of each combination of inspection and monitoring provides its own Pareto front. From the Pareto fronts of the possible cases of combined inspections and monitorings, the final Pareto front can be found, as shown in Figure 6.7. Figure 6.8(a) shows the final Pareto front of the bi-objective optimization problem when the number of inspections and/or monitorings is at most five. The final Pareto front provides the number of inspections and monitorings, sequence of inspections and monitorings, inspection and monitoring application times, and monitoring duration. The optimum inspection and monitoring plans associated with the solutions $P_{c,b1}$, $P_{c,b2}$, $P_{c,b3}$ and $P_{c,b4}$ in Figure 6.8(a) are illustrated in Figure 6.8(b).

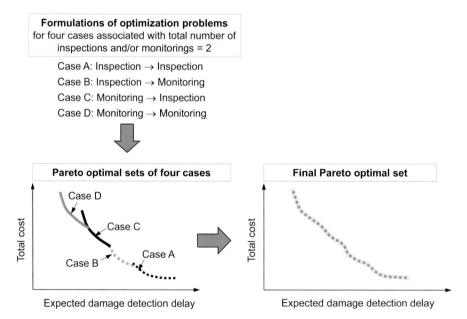

Figure 6.7   Pareto fronts of four combined inspections and monitorings, and final Pareto front (adapted from Kim and Frangopol 2012).

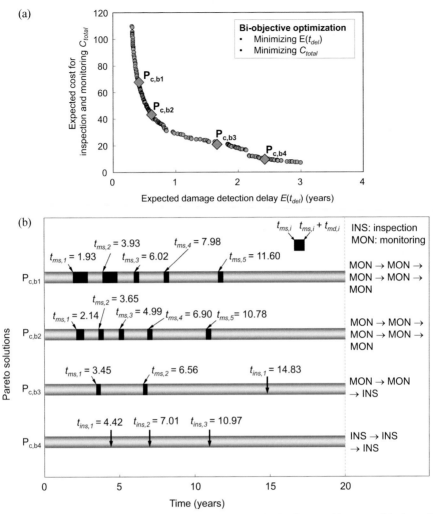

Figure 6.8 (a) Pareto front of the bi-objective optimization problem considering all combinations of inspection and monitoring; and (b) optimum inspection and monitoring plans of solutions $P_{c,b1}$, $P_{c,b2}$, $P_{c,b3}$ and $P_{c,b4}$ (adapted from Kim and Frangopol 2012).

6.3.4 Optimum Inspection and Monitoring Planning with Updating

As mentioned previously, using the information from each inspection and monitoring, the inspection and monitoring plans can be updated with bridge performance deterioration prediction. This updating process is necessary to improve the accuracy and reliability of the inspection and monitoring planning. The schematic of the computational platform for optimum inspection and monitoring planning with updating is illustrated in Figure 6.9. The effect of updating on the optimum inspection and monitoring planning of RC bridges based on minimizing the expected damage detection delay was investigated by Kim and Frangopol (2011c). Figure 6.10(a) shows the PDFs of corrosion initiation time of the RC bridge E-17-HS located in Colorado before and after updating. The comparison between the expected damage detection delays associated with the corrosion initiation times before and after updating is provided in Figure 6.10(b). The expected damage detection delay in Figure 6.10(b) is the result of an optimization to minimize its value. It is shown that updating can substantially affect the expected damage detection delay. Soliman and Frangopol (2014) introduced the updating process of optimum inspection planning, where a single objective of minimizing the total expected cost consisting of inspection cost and expected failure cost is applied to determine the optimum inspection times. Their study shows that the parameters involved in fatigue crack propagation model can be updated with the detected fatigue crack by using the Markov chain Monte Carlo (MCMC) sampling technique, and the remaining service life prediction and optimum inspection time can be updated. A probabilistic framework that utilizes inspection information for optimum decision making regarding the choice and time of inspection actions was proposed by Liu and Frangopol (2019). The optimum inspection plan is expected to provide the highest utility to the decision maker.

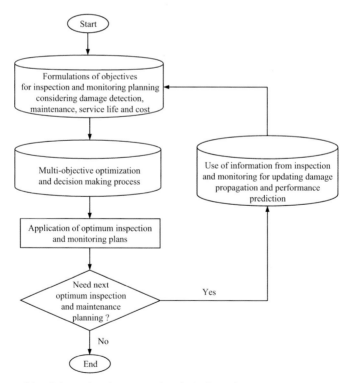

Figure 6.9 Schematic of computational platform for optimum inspection and monitoring planning with updating.

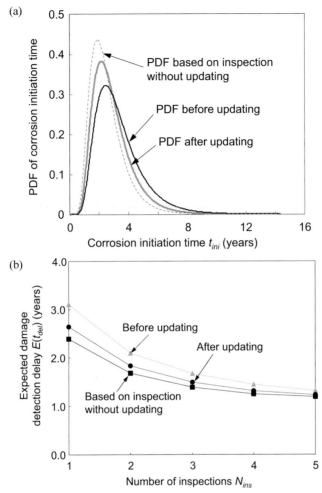

Figure 6.10 Updating effects on (a) PDF of corrosion initiation time; and (b) minimum expected damage detection delay (adapted from Kim and Frangopol 2011c).

6.4   PROBABILISTIC OPTIMUM BRIDGE MAINTENANCE PLANNING

As described in Section 5.3, the objectives associated with optimum bridge life-cycle management planning can be categorized into: (a) performance-based objectives, (b) cost-based objectives, (c) damage detection-based objectives, (d) service life-based objectives, and (e) risk-based objectives. These objectives can be used separately or simultaneously for optimum bridge maintenance planning. Several associated applications investigated over the last few decades are presented in this section. Moreover, the optimum bridge maintenance planning based on the information from inspection and monitoring is illustrated.

6.4.1 Optimum Bridge Maintenance Planning with Single Objective

The total life-cycle cost C_{life} has been extensively used for optimum maintenance planning of civil and marine structures subjected to aging and deterioration mechanisms (Frangopol 2011; Frangopol and Soliman 2016; Frangopol and Kim 2019). The general concept of the life-cycle cost C_{life} is provided in Section 1.2.2. The formulation of C_{life} and the relation among the costs involved in C_{life} are described in Section 5.3.2.

The optimization based on minimizing the expected life-cycle cost while maintaining a minimum target lifetime reliability of RC bridges under corrosion was introduced by Frangopol et al. (1997b). The proposed optimization approach incorporates (a) the quality of inspection techniques with different detection capabilities; (b) all repair possibilities based on an event tree; (c) the effects of aging, deterioration, and subsequent repair on structural reliability; and (d) the time value of money. The overall cost to be minimized includes the initial cost and the costs of preventive maintenance, inspection, repair, and failure. An interior T-girder of a prefabricated RC bridge was used to demonstrate the application of the proposed optimization approach.

Figure 6.11(a) shows the comparison between the life-cycle costs for the optimum uniform and non-uniform time interval inspections (denoted as $C^*_{life,u}$ and $C^*_{life,nu}$, respectively). The effect of inspection quality on optimum number of inspections and repairs is illustrated in Figure 6.11(b). The minimum life-cycle costs for low and high corrosion rates (denoted as $C^*_{life,lc}$ and $C^*_{life,hc}$, respectively) are compared in Figure 6.11(c). The effect of failure cost on optimum number of inspections and repairs is shown in Figure 6.11(d). Several significant findings presented by Frangopol et al. (1997b) are presented such as (a) optimum non-uniform time interval inspection and repair planning results in less life-cycle cost than optimum uniform time interval inspection and repair planning (i.e., $C^*_{life,nu}$ $\leq C^*_{life,nu}$ as shown in Figure 6.11(a)); (b) as the quality of an inspection method increases, the optimum number of inspections decreases (see Figure 6.11(b)); (c) an increase in corrosion rate leads to an increase in the optimum number of inspections and repairs (i.e., $N^*_{ins,hc} \geq N^*_{ins,lc}$, where $N^*_{ins,hc}$ and $N^*_{ins,lc}$ are the optimum number of inspections for high and low corrosion rates, respectively), and increase in life-cycle cost (i.e., $C^*_{life,hc} \geq C^*_{life,lc}$), as shown in Figure 6.11(c); and (d) more optimum inspections and repairs are required if the failure cost increases (see Figure 6.11(d)). This pioneering investigation has served as a basis for more advanced life-cycle cost analyses.

A system reliability-based approach for optimizing the lifetime repair strategy for highway bridges including all their components (i.e., deck, girders, pier caps, expansion bearings, columns, and footing) was presented by Frangopol and Estes (1997) and Estes and Frangopol (1999). The limit states for flexure, shear, and crushing of the columns and expansion bearings are adopted to compute the component reliability. Through a series-parallel modeling with the failure modes of individual components, the time-dependent system reliability is computed to estimate the total lifetime repair costs and the expected life of the entire bridge

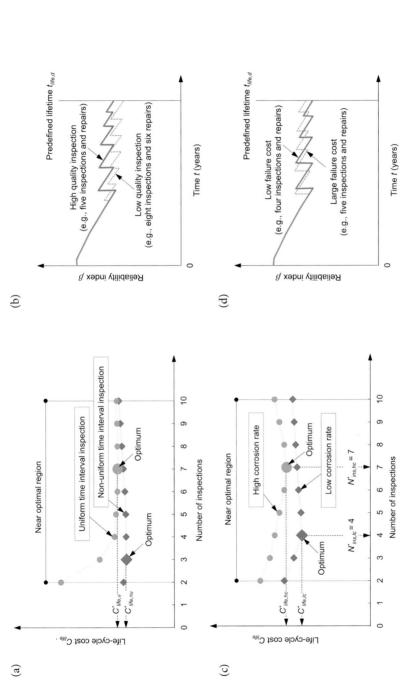

Figure 6.11 Life-cycle costs and reliability index profiles: (a) comparison between life-cycle costs associated with optimum uniform and non-uniform time interval inspections; (b) effect of inspection quality on optimum number of inspections and repairs; (c) comparison between life-cycle costs associated with optimum inspections for high and low corrosion rates; and (d) effect of failure cost on optimum number of inspections and repairs (adapted from Frangopol et al. 1997b).

system for all feasible combinations of the given repair options. The optimum repair strategies result in the minimum total lifetime repair cost among the given combinations of the repair options. The effects of bridge system modelings, deterioration rate, and discount rate of money on the optimum repair strategies and minimum total lifetime repair cost were investigated for Colorado Bridge E-17-AH.

Maintenance planning for deteriorating RC bridges was investigated by Enright and Frangopol (1999b), where the optimum inspection and repair times are computed through a single objective of minimizing the expected life-cycle cost for given reliability constraints. The effect of repair on the cumulative-time failure probability of a bridge system was implemented in the formulation of the expected life-cycle cost. The computation of the cumulative-time failure probability is based on the investigations reported by Enright and Frangopol (1998a, 1998b, 1998c, 1999c, 2000). The study of the RC Colorado Bridge L-18-BG (Enright and Frangopol, 1999b) presents two main conclusions: (a) multiple repairs appear to have little influence on the system failure probability for the shotcrete repair option; In fact, for shotcrete repair, a large number of repairs may be required to keep the lifetime system failure probability below a reasonable level, whereas only a single replacement is required to achieve the same goal; and (b) failure cost and discount rate of money can have a significant impact on optimum maintenance planning.

Based on the investigations by Frangopol et al. (2001), Kong (2001), and Kong and Frangopol (2002), a computational program LCADS (Life-Cycle Analysis of Deteriorating Structures) to analyze the life-cycle reliability of deteriorating individual bridges and bridge groups was developed by Kong and Frangopol (2003b). The reliability index profile and related cost function for a deteriorating bridge are implemented in the program. This program uses random variables to compute the reliability index profile and the cost function of deteriorating individual bridges or bridge groups. Also, it can solve a single optimization problem to minimize the expected cumulative maintenance cost during a given service life. As a result, the mean application times of PM and EM can be optimized as shown in Kong and Frangopol (2005).

The effects of limit state selection on optimum repair strategies and associated life-cycle costs can be found in Stewart et al. (2004), where the ultimate strength and serviceability-based limit states are considered. The formulation of the ultimate strength limit state is associated with the flexural failure of an RC bridge deck. The time to reach severe cracking and spalling of the bridge deck is used to define the serviceability limit state. The optimal repair strategies are associated with the minimum present cost of deck replacement. In this investigation, it is shown that the life-cycle cost based on a serviceability limit state can be larger than that based on a strength limit state. Therefore, life-cycle cost analyses should consider multiple limit states, including serviceability and ultimate limit states.

Optimum maintenance planning based on lifetime distributions was introduced and applied by Yang et al. (2006a). The optimum maintenance planning includes (a) establishing the survival functions for all bridge components and system model,

(b) determining all possible maintenance interventions and estimating the related costs; (c) determining all the combinations of possible maintenance interventions; (d) computing the bridge system-level survivor function for each maintenance plan; (e) computing the life-cycle maintenance cost; and (f) finding the optimum maintenance types and times to minimize the life-cycle maintenance cost. Their study shows that the system modeling affects the optimum maintenance strategy and the present value of life-cycle maintenance. Further development can be found in Yang et al. (2006b), where both total cumulative maintenance cost and expected failure cost during a predefined lifetime are minimized to find the optimum maintenance interventions. The investigation by Yang et al. (2006b) reveals that (a) a larger maintenance cost results in the optimum maintenance interventions associated with less frequent application of maintenance, and (b) there is a significant dependence among the optimum maintenance interventions, the associated minimum expected total cost, the ratio between maintenance cost and failure cost, and the discount rate of money.

Based on the approaches developed by Yang et al. (2006a), the optimum maintenance planning based on lifetime redundancy was introduced by Okasha and Frangopol (2010a). The formulations of the lifetime availability and redundancy with essential maintenance actions are presented. The effects of a threshold of redundancy on optimum maintenance plan, maintenance cost, and several performance indicators such as probability of failure and cumulative failure rate were investigated. It was shown that a higher redundancy threshold results in more frequent essential maintenance, less probability of failure, less cumulative failure rate, and larger maintenance cost during a predefined lifetime.

6.4.2 Optimum Bridge Maintenance Planning with Multiple Objectives

In general, life-cycle cost and performance interact in a conflicting manner (Frangopol 2011). By considering two competing objectives (e.g., both minimizing the life-cycle cost and maximizing the bridge performance) simultaneously, more rational and flexible bridge maintenance management can be achieved. For this reason, the approaches using a MOOP have been extensively developed and applied for bridge maintenance planning over the last few decades.

Liu and Frangopol (2005b) presented the application of a deterministic MOOP maintenance planning considering three objectives: minimizing the condition index, maximizing the safety index, and minimizing the life-cycle maintenance cost of a deteriorating RC bridge. The condition index is defined by discrete values on a scale varying from zero (i.e., best condition) to three (i.e., worst condition), and the safety index is quantified by the ratio of available to required live-load carrying capacity. The design variables of the MOOP are the times of first and subsequent maintenance (i.e., silane treatment) applications. A genetic algorithm is used to find the Pareto front. Figure 6.12 shows this Pareto front in 3D Cartesian coordinates and its 2D projections. The effects of the uncertainties associated with deterioration models on condition, safety and cost profiles were investigated.

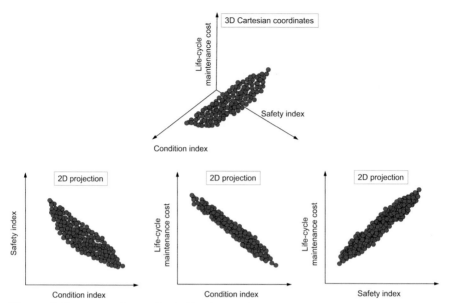

Figure 6.12   Pareto front of tri-objective optimization considering minimizing the condition index, maximizing the safety index, and minimizing the life-cycle maintenance cost (adapted from Liu and Frangopol 2005b).

Furuta et al. (2006) used a deterministic MOOP for bridge maintenance planning for deteriorating multiple bridge components, where three objectives are used: (a) minimizing the life-cycle cost; (b) maximizing the safety level; and (c) maximizing the service life. In the formulation of the optimization problem, eight types of maintenances (i.e., surface painting, surface covering, section restoring, desalting, cathodic protection, section restoring with surface, covering and reconstruction) and six bridge components (i.e., two piers, shoe, steel girder, and two RC slabs) are considered. The Pareto front of the MOOP indicate the times and types of maintenance for each of the six bridge components. Their study shows that an increase in safety level and/or service life produces an increase in a life-cycle cost.

A probabilistic MOOP maintenance planning, considering three objectives including minimizing the worst condition index, maximizing the worst safety index, and minimizing the life-cycle maintenance cost for a given lifetime, was applied in the investigation of Liu and Frangopol (2005d). The probabilistic condition index and safety index profiles with four types of maintenance (i.e., replacement of expansion joint, silane, cathodic protection, minor concrete repair) are adopted for a deteriorating RC bridge. The probabilistic parameter data to represent these profiles are provided by Denton (2002). A genetic algorithm is used to find the Pareto front, which is associated with annualized maintenance interventions.

A further probabilistic MOOP for bridge maintenance planning was investigated by Neves et al. (2006a, 2006b). In this optimization, the condition index, safety index, and cumulative maintenance cost are used as objective functions. These indices and cost are represented by the multi-linear profiles developed in Neves and Frangopol (2005). Neves et al. (2006a, 2006b) use a full probabilistic MOOP, which treats the parameters involved in the multi-linear profiles as random variables. In this MOOP, the safety index threshold and the mean times of first and subsequent maintenance actions are optimized. The optimum bridge maintenance planning of Neves et al. (2006a) is based on a single maintenance type (i.e., only silane or rebuild). The combination of maintenance types (i.e., both silane and rebuild) for the formulation of the MOOP maintenance planning is reported by Neves et al. (2006b). From the investigations of Neves et al. (2006a, 2006b), it is concluded that the combinations of different maintenance actions can be more cost-effective than single maintenance actions. The developments of life-cycle maintenance planning for deteriorating civil infrastructures using multiple objectives related to the condition, safety and life-cycle cost are reviewed in Frangopol and Liu (2007b).

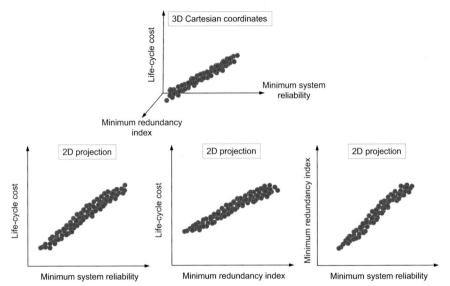

Figure 6.13 Pareto front of tri-objective optimization considering maximizing the minimum system reliability, maximizing the minimum redundancy index and minimizing the life-cycle cost (adapted from Okasha and Frangopol 2009b).

Okasha and Frangopol (2009b) dealt with a MOOP considering maximizing the system reliability, maximizing the redundancy and minimizing the life-cycle cost. The formulation of MOOP, including system reliability analysis and cost estimation with maintenance, is based on the outcomes of Estes and Frangopol (1999) and Akgül and Frangopol (2004d). The MOOPs associated with the two maintenance strategies are compared. The first maintenance strategy assumes that multiple maintenances are applied to different components at the same time. The second maintenance strategy is based on the premise that multiple maintenance actions are applied to different components at different times. The optimum maintenance planning obtained from the bi- and tri-objective optimizations are also compared. Figure 6.13 shows the Pareto front of the tri-objective optimization in 3D Cartesian coordinates and its projections in 2D coordinates. Their study found that an increase in the system reliability may worsen the system redundancy, and the improvement of structural performance may not lead to a proportional increase in life-cycle cost.

The probabilistic MOOP approach for bridge maintenance planning was used by Okasha and Frangopol (2010b), in which the uniform and non-uniform time interval of PM, the times and types of combined PM and EM types are all optimized. The objective set of the MOOP consists of minimizing the worst value of unavailability, maximizing the worst value of redundancy index, and minimizing the life-cycle maintenance cost. The unavailability and redundancy index with PM and EM are formulated using lifetime functions.

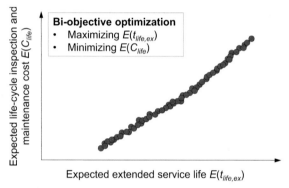

Figure 6.14   Pareto front of the bi-objective optimization considering maximizing the expected service life and minimizing the expected life-cycle inspection and maintenance cost (adapted from Kim et al. 2013).

The generalized framework for MOOP inspection and maintenance planning was developed by Kim et al. (2013). It can be applied to any type of deteriorating structures. This framework consists of several parts: (a) prediction of damage occurrence and propagation and service life of a deteriorating structure under uncertainty; (b) determining the relation between the degree of damage and the probability of damage detection of an inspection method; and (c) formulation of the service life considering effects of inspection and maintenance on the service life and life-cycle cost under uncertainty. The proposed approach was applied to the RC slab of the Colorado Bridge E-16-Q. The MOOP of this framework is based on two conflicting objectives: maximizing the expected extended service life and minimizing the expected life-cycle inspection and maintenance cost. The probabilistic formulations of these objectives use the decision tree representing all the possible outcomes and associated occurrence probability related to damage detection and maintenance applications under uncertainty. The Pareto front of the MOOP is shown in Figure 6.14. The effects of the probability of damage detection of an inspection method, number of inspections, and damage criteria to select maintenance types on the expected extended service life, and expected life-cycle inspection and maintenance costs are addressed.

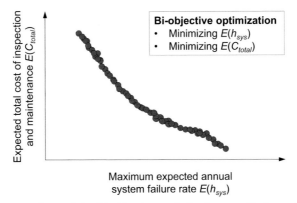

Figure 6.15   Pareto front of the bi-objective optimization considering minimizing both the maximum expected system failure rate and the expected total cost of inspection and maintenance (adapted from Barone et al. 2014).

The application of MOOP for inspection and maintenance planning presented by Barone et al. (2014) considers two conflicting objectives: minimizing both the lifetime maximum expected system failure rate and the expected total cost of inspection and maintenance. The proposed inspection and maintenance planning is system-based, which provides the optimum times for a bridge system inspection and optimum maintenance action for each bridge component. As a case study, the proposed method has been applied to the superstructure (i.e., RC deck and nine steel girders) of the Colorado Bridge E-17-AH. The relation between the maximum expected system failure rate and the expected total cost of the Pareto front are shown in Figure 6.15. It was concluded that: (a) the expected total cost of all inspections and maintenance actions during the lifetime of a structural system is component dependent, whereas the expected system failure rate depends on both the system configuration and component failure rate, and (b) different maintenance strategies can be chosen from the Pareto front. Low-cost maintenance plans are primarily associated with no repair or preventive maintenance, providing a small reduction of the expected system failure rate. In these cases, in-depth inspections should be concentrated in the early life of the structure. Maintenance plans with the highest effect on the structural performance are generally associated with in-depth inspections distributed along the last part of the life-cycle of the system. For these strategies, essential maintenance options on critical components are dominant. Furthermore, Barone and Frangopol (2014b) compares four bi-objective optimum maintenance plannings consisting of life-cycle maintenance cost and four different performance indicators (i.e., reliability index, risk, availability, and hazard). These optimization problems are applied for the determination of the optimal maintenance planning of the superstructure of the RC E-17-HS Bridge located in Colorado. Lifetime distributions are used to formulate these performance indicators. The bi-objective optimization determines the application times and numbers of EM for selected bridge components. The four objectives of minimizing the maximum system risk, minimizing the maximum annual system failure probability, maximizing the minimum system availability, and minimizing the maximum system hazard are taken into account separately with minimizing the life-cycle maintenance cost for bi-objective optimization. It was concluded that: (a) risk-based optimum maintenance plans are more cost-efficient than their reliability-based counterparts, and (b) availability-based and hazard-based optimization formulations may result in similar solutions.

6.4.3   Optimum Bridge Maintenance Planning with Updating Based on Monitoring

As described previously, the information from inspection and monitoring can be used for predicting and updating the bridge performance and service life. Orcesi et al. (2010) presented the MOOP to determine the optimum maintenance application times and types for superstructures of steel girder bridges (i.e., deck and girders) incorporating monitoring results. Monitoring data enables to update the knowledge on structural performance and increases the accuracy of the structural reliability analysis. The two objectives of minimizing both the expected maintenance cost and failure cost are considered, in which the reliability index is estimated for the multiple limit states including fatigue performance, serviceability and ultimate strength. The state functions for fatigue and serviceability are formulated using long-term monitoring, and the associated time-dependent reliability indices are predicted. It is shown that the consideration of multiple limit states can lead to different optimal maintenance solutions, if taken individually or simultaneously.

Further investigation to include monitoring data in the optimal bridge maintenance strategies updating was performed by Orcesi and Frangopol (2011a). The monitoring data of the effective stress ranges and stress cycles of I-39 Northbound Bridge in Wisconsin is used to update the lifetime functions and the time to failure using the Bayesian theorem. The details on the Bayesian theorem are explained in Section 2.2.4. Before updating, the optimum bridge maintenance strategies are searched by using the MOOP consisting of minimizing both the expected maintenance cost and failure cost. When the monitoring is performed, the time to failure is updated, and the optimum bridge maintenance strategies are determined again through the MOOP. Using the same Wisconsin bridge as an example, Orcesi and Frangopol (2011b) used SHM information and statistics of extremes to determine a performance margin at each decision time. The optimal solution is searched by simultaneously minimizing the expected maintenance cost, the expected failure cost, and the expected error of decision. The design variables are the interval between two maintenance decision times, the monitoring duration, and a decision parameter. By introducing the error of the decision process in the optimization procedure, it is possible to quantify how maintenance optimization solutions are affected by monitoring occurrence and duration.

A MOOP-based approach to determine the optimal cut-off area of the retrofitting distortion-induced fatigue cracking of steel bridges was proposed by Liu et al. (2010b). The cut-off area is optimized considering two conflicting objectives: maximizing the fatigue reliability of the connection details after retrofitting and minimizing the cut-off area. The monitored data collected on an existing steel tied-arch bridge (i.e., Birmingham Bridge) in Pennsylvania is used to estimate the fatigue stress ranges and the number of cycles, and to compute the fatigue reliability at the critical locations after retrofitting. The estimated stress range was validated by using finite element analysis (FEA).

Okasha and Frangopol (2012) developed a life-cycle framework for SHM-based optimum bridge maintenance management using system performance concepts. The developed framework consists of (a) bridge performance prediction, (b) bridge maintenance optimization, and (c) updating of bridge performance prediction and maintenance optimization using data from the controlled testing and long-term monitoring. The advanced tools and techniques to predict the bridge performance using a finite element model (FEM) and monitoring were developed by Okasha and Frangopol (2010c) and Okasha et al. (2012). The concepts presented are illustrated on the I-39 Wisconsin Bridge. The superstructure of the bridge is composed of four composite steel plate I-section girders carrying a RC slab. Based on these developed tools and techniques, Okasha and Frangopol (2011) proposed a detailed computational platform to integrate SHM in a bridge life-cycle bridge management framework. The elements integrated into the framework include the advanced assessment of life-cycle performance, analysis of system and component performance interaction, advanced maintenance optimization, and updating the life-cycle performance by information obtained from SHM and controlled testing.

Yang and Frangopol (2021) compare static and adaptive risk-based inspection planning methods. The comparison, based on a generic Markovian deterioration model, demonstrates that adaptivity of inspection plans is able to reduce the total expected life-cycle cost of deteriorating structures.

6.5   CONCLUDING REMARKS

In this chapter, several applications associated with optimum life-cycle bridge management planning are presented with emphasis on probabilistic performance and service life prediction, optimum inspection and monitoring planning, and optimum maintenance planning at an individual bridge-level. The presented applications are selected from previous investigations on optimum life-cycle bridge management planning of an individual bridge, and are summarized to highlight the novelties of the investigations. The life-cycle bridge network maintenance management is addressed in Chapter 7.

Chapter 7

Life-Cycle Bridge Network Management

NOTATIONS (continued)	
N_{sc}	= number of failure scenarios of bridges
OD_{ij}	= component of the origin-destination matrix
$p_{s,i}$	= reliability of the ith bridge
p_s^{net}	= bridge network connectivity reliability
p_{us}	= probability of unsatisfactory performance of bridge network
P_i^{an}	= annual occurrence probability associated with the ith branch of an event tree
$P_{tr,l}$	= occurrence probability of traffic congestion on the link l
r_{dis}	= annual discount rate of money
r_β	= deterioration rate
R_K^{net}	= bridge network risk
RI_{net}	= bridge network redundancy index
t_{ini}	= deterioration initiation time
$t_{life,d}$	= predefined time period
TTD	= total travel distance
TTT	= total travel time
β_{in}	= initial reliability index
β^{net}	= reliability index of bridge network connectivity
$\Omega^{(-)}$	= objective set consisting of objectives to be minimized
$\Omega^{(+)}$	= objective set consisting of objectives to be maximized

ABSTRACT

In Chapter 7, probabilistic life-cycle bridge network management is presented. The probabilistic performance indicators used for bridge network management are: (a) connectivity-based; (b) travel flow-based; (c) cost-based; and (d) risk-based. The time-dependent bridge network performance prediction with maintenance interventions is described. The effects of maintenance interventions applied to individual bridges on the bridge network performance are investigated. The general concepts and applications of the multi-objective optimum life-cycle bridge network management are described. The life-cycle bridge network management presented in this chapter is based on the bridge performance deterioration under normal loading effects and environmental conditions.

7.1 INTRODUCTION

The scale of the activity of managing bridges has increased over the last four decades from managing a few individual bridges to hundreds and thousands of bridges located in widespread geographical areas. Life-cycle management at bridge network level considering the minimization of life-cycle cost while preserving a minimum level of bridge performance leads to an improvement in the allocation of limited financial resources (Augusti et al. 1998; Bocchini and Frangopol 2011a, 2011b; Hu and Madanat 2015; Hu et al. 2015; Yang and Frangopol 2018b). In order to achieve a successful bridge network management under uncertainty, a proper understanding of the probabilistic time-dependent network performance analysis and effects of maintenance interventions on the bridge network performance and life-cycle cost is required. The optimization process based on the bridge network performance and life-cycle cost is essential to allocate limited resources in an efficient way to balance the cost and performance, and determine the best maintenance interventions for individual bridges in a network (Furuta et al. 2011; Frangopol and Bocchini 2012; Biondini and Frangopol 2015).

In this chapter, the life-cycle bridge network management approach under uncertainty is presented. The probabilistic performance indicators used for bridge network management are provided. These indicators are based on connectivity, travel flow, cost, and risk. The connectivity-based network performance indicators are represented by the bridge network connectivity reliability, connectivity level, and redundancy index. The total travel time, distance, and cost are involved in the travel flow-based network performance indicator. The cost-based network performance indicators include the expected maintenance cost, user cost, life-cycle failure cost, and total indirect cost. The time-dependent bridge network performance prediction with maintenance interventions is described and the effects of time-dependent performance of individual bridges with maintenance interventions on the bridge network performance are addressed. The general concepts and applications of the multi-objective optimum life-cycle bridge network management are described. The life-cycle bridge network management presented in this chapter is based on the performance deterioration of individual bridges under normal loading effects and environmental conditions.

7.2   PROBABILISTIC BRIDGE NETWORK PERFORMANCE

The bridge network performance is affected by the performance of individual bridges (Orcesi and Cremona 2011; Dong and Frangopol 2019). Due to the uncertainty associated with the performance of individual bridges, the bridge network performance should be estimated in a probabilistic way (Bocchini and Frangopol 2011a). In general, bridge networks can be modeled using a series, parallel, or series-parallel system consisting of individual bridges (Liu and Frangopol 2005a; Frangopol and Kim 2011). The formulation of a bridge network consisting of individual bridges may be similar to the formulation of a bridge system with its components (e.g., bridge deck, girders, and piers). However, the individual bridges in a network are usually built in different years, with different materials, and using different design specifications. Over the years, these bridges have been subjected to different deterioration mechanisms, loading and environmental conditions (Akgül and Frangopol 2004a, 2004d, 2004e; Liu and Frangopol 2005c, 2006a). For this reason, the bridge network performance cannot be rationally quantified using only the performance indicators presented in Section 1.4. The bridge network performance should consider reliability, connectivity, travel time and distance, maintenance cost, user cost, failure cost, risk, sustainability, and resilience. The network performance can be estimated based on the performance of individual bridges in the network (Frangopol and Bocchini 2012; Dong and Frangopol 2019). The performance of bridge networks can be rationally quantified based on (a) connectivity; (b) travel flow; (c) cost; and (d) risk. The performance indicators involved in these four categories are summarized in Figure 7.1.

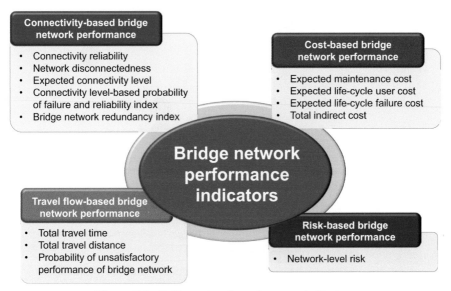

Figure 7.1 Bridge network performance indicators.

7.2.1 Connectivity-Based Bridge Network Performance

A bridge network can be modeled by using nodes and links. The nodes correspond to the locations of interest such as cities in a statewide network and hospitals in a small regional network. The links are represented by roads including bridges (Quimpo and Wu 1997; Liu and Frangopol 2005c). The bridge network connectivity reliability indicates the probability that traffic can reach the destination from the origin (Bell and Iida 1997; Liu and Frangopol 2005a). Figure 7.2 shows the block diagram and event tree for a schematic bridge network consisting of three bridges and two nodes, where the two nodes O and D indicate the origin and destination, respectively. Using the event tree in Figure 7.2(b), the bridge network connectivity reliability p_s^{net} is expressed as (Liu and Frangopol 2005a)

$$p_s^{net} = \sum_{i=1}^{N_{cn}} \left(\prod_{j \in \mathbf{S}}^{N_{s,i}} (p_{s,j}) \cdot \prod_{k \in \mathbf{F}}^{N_{f,i}} (1 - p_{s,k}) \right) \quad (7.1)$$

where N_{cn} is the number of branches of event tree representing the connection, $N_{s,i}$ and $N_{f,i}$ are the number of bridges involved in the sets **S** and **F**, respectively, **S** and **F** indicate the bridge sets associated with the safe and failure states, respectively, and $p_{s,j}$ is the reliability of the *j*th bridge. Equation (7.1) is based on the assumption that the states of the individual bridges in the network are statistically independent. The bridge network connectivity reliability considering the correlation among the individual bridges can be computed using several software programs (see Section 1.3.2), where the series-parallel system in Figure 7.2(c) can be applied for the bridge network shown in Figure 7.2(a).

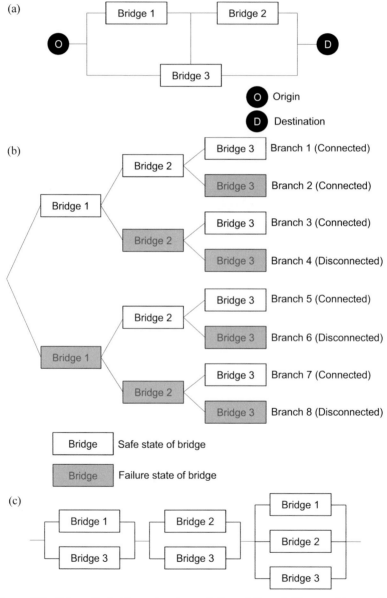

Figure 7.2   Schematic bridge network consisting of three bridges and two nodes: (a) block diagram; (b) event tree; (c) series-parallel model.

In order to quantify the network disconnection (i.e., not reaching a node from every other node), the network disconnectedness D_{net} is defined as (Ng and Efstathiou 2006)

$$D_{net} = \frac{N_{up}}{N_{link}} \qquad (7.2)$$

where N_{up} is the number of unreachable pairs of nodes, and N_{link} is the maximum possible number of links in the network. The bridge network performance to measure the possibility to reach every node of the bridge network can be represented by the connectivity level CL as (Bocchini and Frangopol 2013)

$$CL = \sum_{i=1}^{N_{nd}} \sum_{j=1}^{N_{nd}} L_{ij} \times OD_{ij} \qquad (7.3)$$

where N_{nd} is the number of nodes in the bridge network, L_{ij} is the state value to show the possibility that the route from the node i to node j is in-service (i.e., $L_{ij} = 1.0$ for in-service and 0.0 for out of service), and OD_{ij} is the component of the origin-destination matrix **OD** formulated based on the number of vehicles traveling from the node i to node j during a given time interval. The normalized connectivity level CL^{norm} is expressed as

$$CL^{norm} = \frac{CL - CL_0}{CL_{all} - CL_0} \qquad (7.4)$$

where CL_{all} and CL_0 are the connectivity levels when all the bridges are in service and out of service, respectively. Considering the uncertainties associated with L_{ij} and OD_{ij} in Eq. (7.3), the expected normalized connectivity level $E(CL^{norm})$ can be used for bridge network performance (Bocchini and Frangopol 2011a, 2013).

Using the normalized connectivity level CL^{norm} of Eq. (7.4), the state function g_{CL} can be expressed as

$$g_{CL} = CL^{norm} - CL_{th} \qquad (7.5)$$

where CL_{th} is the threshold of the normalized connectivity level. Based on the state function g_{CL} of Eq. (7.5), the probability of failure $p_{f,CL}^{net}$ and reliability index β_{CL}^{net} are expressed, respectively, as (Bocchini and Frangopol 2013)

$$p_{f,CL}^{net} = P(g_{CL} < 0) \qquad (7.6a)$$

$$\beta_{CL}^{net} = \Phi^{-1}(1 - p_{f,CL}^{net}) \qquad (7.6b)$$

where Φ^{-1} = inverse of the standard normal cumulative distribution.

The concept of structural redundancy described in Section 1.4 can also be used as a bridge network performance indicator of the warning before the network disconnection. The redundancy index RI_{net} of the bridge network is expressed as (Kim et al. 2020)

$$RI_{net} = \sum_{i=1}^{N_{bdg}} r_{oc,i} \cdot RI_{net,i} \tag{7.7}$$

where N_{bdg} is the total number of individual bridges in the network, and $r_{oc,i}$ and $RI_{net,i}$ are the normalized occurrence rate and bridge network redundancy index, respectively, when the ith individual bridge in a network is in a failure state. The normalized occurrence rate $r_{oc,i}$ in Eq. (7.7) is

$$r_{oc,i} = \frac{1 - p_{s,i}}{\sum_{j=1}^{N_{bdg}} (1 - p_{s,j})} \tag{7.8}$$

The bridge network redundancy index $RI_{net,i}$ in Eq. (7.7) is (Frangopol and Curley 1987; Saydam and Frangopol 2011)

$$RI_{net,i} = \frac{\beta^{net}}{\beta^{net} - \beta_i^{net}} \tag{7.9}$$

where β^{net} is the reliability index of the network based on the bridge network connectivity reliability p_s^{net} defined in Eq. (7.1). The relation between p_s^{net} and β^{net} is indicated in Eq. (1.10). β_i^{net} is the reliability index of a network when the ith individual bridge in the network is in a failure state. As shown in Figure 7.1, the bridge network performance based on connectivity can be represented by the bridge network connectivity reliability p_s^{net}, network disconnectedness D_{net}, expected normalized connectivity level $E(CL^{norm})$, connectivity level-based probability of failure $p_{f,CL}^{net}$ and reliability index β_{CL}^{net}, and bridge network redundancy index RI_{net}.

7.2.2   Travel Flow-Based Bridge Network Performance

The travel flow has been generally adopted to evaluate the bridge network performance, quantified by total travel time, distance, and user satisfaction (Liu and Frangopol 2006a; Frangopol and Bocchini 2012; Dong and Frangopol 2019). The total travel time TTT is a summation of the required time to reach every node from every other node during a fixed time window. Accordingly, the total travel time TTT_I can be expressed as (Scott et al. 2006)

$$TTT_I = \sum_{i \in \mathbf{I}}^{N_{nd}} \sum_{j \in \mathbf{J}}^{N_{ij}} f_{ij} t_{ij} \tag{7.10}$$

where i and j = nodes of the bridge network; $\mathbf{I}$ = set of network nodes; N_{nd} = total number of nodes of $\mathbf{I}$; $\mathbf{J}$ = subset of nodes that can be reached from the node i; and N_{ij} = total number of nodes of $\mathbf{J}$. f_{ij} and t_{ij} are the traffic flow and travel time associated with the link ij, respectively. Considering the effect of the traffic flow f_{ij} on travel time t_{ij}, *TTT* of Eq. (7.10) can be modified as (Bocchini and Frangopol 2010)

$$TTT_{II} = \sum_{i \in \mathbf{I}}^{N_{nd}} \sum_{j \in \mathbf{J}}^{N_i} \int_0^{f_{ij}} \tau_{ij}(f) df \qquad (7.11)$$

where $\tau_{ij}(f)$ is the travel time function of traffic flow f for the link ij. Furthermore, the total travel distance *TTD* is estimated as (Frangopol and Bocchini 2012)

$$TTD = \sum_{i \in \mathbf{I}}^{N_{nd}} \sum_{j \in \mathbf{J}}^{N_i} f_{ij} d_{ij} \qquad (7.12)$$

where d_{ij} is the distance between nodes i and j.

The bridge network performance can be represented by user satisfaction. Liu and Frangopol (2006a) introduced the probability of unsatisfactory performance of bridge network p_{us} as

$$p_{us} = \frac{\displaystyle\sum_{l=1}^{N_l} f_l \times c_{tr,l} \times P_{tr,l}}{\displaystyle\sum_{l=1}^{N_l} f_l \times c_{tr,l}} \qquad (7.13)$$

where N_l = total number of links; f_l = traffic flow on the link l; $c_{tr,l}$ = unit travel costs; and $P_{tr,l}$ = occurrence probability of traffic congestion on the link l. The unit travel cost $c_{tr,l}$ is expressed as (Mackie and Nellthorp 2001)

$$c_{tr,l} = \alpha t_l + \gamma d_l \qquad (7.14)$$

where t_l and d_l are the travel time and distance for the link l, respectively. α is the unit cost for travel time, and γ is the unit cost for travel distance. In order to compute the occurrence probability $P_{tr,l}$ in Eq. (7.13), the state function g_{tr} in terms of the traffic flow is formulated as (Liu and Frangopol 2006a)

$$g_{tr} = f_l^c - f_l \qquad (7.15)$$

where f_l^c is the capacity of traffic flow on the link l. When f_l^c and f_l are treated as random variables, the occurrence probability $P_{tr,l}$ corresponds to the probability that f_l^c is less than f_l (i.e., $g_{tr} < 0$). The capacity of traffic flow f_l^c is affected by the free-flow traffic speed and number of lanes. The link traffic flow f_l can be computed using the annual average daily traffic (AADT) of the link. The detailed computations of f_l^c and f_l can be found in TRB (2000) and Liu and Frangopol (2006a). As shown in Figure 7.1, the travel flow-based bridge network performance indicator includes the total travel time TTT_I and TTT_{II}, total travel distance *TTD*, and the probability of unsatisfactory performance of bridge network p_{us}.

7.2.3   Cost-Based Bridge Network Performance

The cost-based bridge network performance includes the life-cycle bridge network maintenance cost, user cost and failure cost (Liu and Frangopol 2006a). The expected maintenance cost of a bridge network C_{ma}^{net} is the total cost to maintain all the bridges in the network during a predefined time period $t_{life,d}$ as (Liu and Frangopol 2005a, 2006b; Bocchini and Frangopol 2011b; Kim et al. 2020)

$$ C_{ma}^{net} = \sum_{i=1}^{N_{bdg}} \sum_{j=1}^{N_{ma,i}} \frac{C_{ma,ij}}{\left(1+r_{dis}\right)^{t_{ij}}} \tag{7.16} $$

where N_{bdg} is the total number of individual bridges in the network, $N_{ma,i}$ is number of maintenance actions for the ith individual bridge, r_{dis} is the discount rate of money, and t_{ij} and $C_{ma,ij}$ are the application time and cost of the jth maintenance for the ith individual bridge during the predefined time period $t_{life,d}$.

The user cost associated with a bridge network is calculated as the cost for the AADT to travel within the bridge network, which considers costs associated with travel time and distance, and monetary loss induced by bridge network failure (Daniels et al. 1999; Mackie and Nellthorp 2001; Thoft-Christensen 2009). Using the event tree in Figure 7.2(b), the expected life-cycle user cost $C_{us,life}^{net}$ during a predefined time period $t_{life,d}$ can be estimated as (Liu and Frangopol 2006b)

$$ C_{us,life}^{net} = \sum_{i=1}^{N_{br}} \sum_{j=1}^{N_t} \frac{C_{us,i}^{net}\left(j\right)}{\left(1+r_{dis}\right)^{j}} \cdot P_i^{an}\left(j\right) \tag{7.17} $$

where N_{br} is the number of branches of the event tree, N_t is the number of years in the time period $t_{life,d}$, and $C_{us,i}^{net}(j)$ and $P_i^{an}(j)$ are the user cost and annual occurrence probability associated with the ith branch of the event tree in the jth year. The user cost $C_{us,i}^{net}(j)$ can be computed as

$$ C_{us,i}^{net}\left(j\right) = \left\{ \sum_{m=1}^{N_l} \left[\left(c_{tr,m} + c_{fail,i,m}\right) AADT_m\left(j\right)\right] \right\} \cdot T_{re,i}\left(j\right) \tag{7.18} $$

where $c_{tr,m}$ and $c_{fail,i,m}$ are the unit travel costs associated with the ith branch when the mth link is intact and failed, respectively. $AADT_m(j)$ is the annual average daily traffic volume on mth link in jth year. $T_{re,i}(j)$ is the time period required to restore to the full network functionality from the ith branch in jth year. If the states of the individual bridges in the network are statistically independent, $P_i^{an}(j)$ in Eq. (7.17) is estimated as

$$ P_i^{an}\left(j\right) = \prod_{k \in S}^{N_{s,i}} \left[p_{s,k}\left(j\right)\right] \cdot \prod_{l \in F}^{N_{f,i}} \left[1 - p_{s,l}\left(j\right)\right] \tag{7.19} $$

where $N_{s,i}$ and $N_{f,i}$ are the number of bridges involved in the safe and failure states, respectively, and $p_{s,k}(j)$ is the reliability of kth bridge in the jth year.

The sum of the number of bridges involved in the safe and failure states is equal to the total number of individual bridges in the network (i.e., $N_{bdg} = N_{s,i} + N_{f,i}$).

The failure cost of bridge network can indicate the performance for bridge network maintenance management (Kong and Frangopol 2005). The expected life-cycle failure cost of bridge network C_{fail}^{net} for a predefined time period $t_{life,d}$ is expressed as (Liu and Frangopol 2006b)

$$C_{fail}^{net} = \sum_{i=1}^{N_{bdg}} C_{fail,i} \qquad (7.20)$$

where $C_{fail,i}$ is the expected failure cost of the ith individual bridge for a predefined time period $t_{life,d}$. The expected failure cost $C_{fail,i}$ is

$$C_{fail,i} = \int_0^{t_{life,d}} \frac{C_{f,i}(t)}{\left(1 + r_{dis}\right)^t} \cdot f_{tf,i}(t)\,dt \qquad (7.21)$$

where $C_{f,i}(t)$ = expected monetary loss induced by the ith bridge failure at time t, and $f_{tf,i}(t)$ is the PDF of time to failure of the ith bridge.

The indirect cost induced by bridge failure can be included in the cost-based bridge network performance. Considering that bridge failure can produce additional travel time and distance, the total indirect cost of bridge failure C_{ind}^{net} is expressed as (Saydam et al. 2013a)

$$C_{ind}^{net} = C_{TTT}^{net} + C_{TTD}^{net} \qquad (7.22)$$

where C_{TTT}^{net} and C_{TTD}^{net} are the total additional costs due to the increase in *TTT* (see Eqs. 7.10 and 7.11) and *TTD* (see Eq. 7.12), respectively. The total additional costs C_{TTT}^{net} and C_{TTD}^{net} are expressed as

$$C_{TTT}^{net} = c_{tr,t} \cdot t_d \sum_{i=1}^{k} \Delta TTT_i \cdot \frac{t_i}{t_{unit}} \qquad (7.23a)$$

$$C_{TTD}^{net} = c_{tr,d} \cdot t_d \sum_{i=1}^{k} \Delta TTD_i \cdot \frac{t_i}{t_{unit}} \qquad (7.23b)$$

where $c_{tr,t}$ = cost of additional time for the travelers per unit time; t_d = duration of the time step; k = total number of intervals dividing one day with the time duration t_i; t_{unit} = duration of one day; ΔTTT_i = change in total travel time associated with the ith interval of one day; $c_{tr,d}$ = cost of additional time for the travelers per unit distance; and ΔTTD_i = change in total travel distance associated with the ith interval of one day. ΔTTT_i and ΔTTD_i are affected by the failures of bridges of the network. In summary, as shown in Figure 7.1, the expected maintenance cost C_{ma}^{net}, expected life-cycle user cost $C_{us,life}^{net}$, expected life-cycle failure cost C_{fail}^{net}, and total indirect cost C_{ind}^{net} can be used to indicate the bridge network performance.

7.2.4   Risk-Based Bridge Network Performance

In order to address the monetary losses resulted from a bridge failure event, the risk is used as a structural performance indicator, as described in Section 1.4. Similarly, the risk can represent the bridge network performance, where its assessment is based on the quantification of the expected value of losses due to single or multiple bridge failures (Saydam et al. 2013a). The general formulation of the bridge network risk R_K^{net} is (Yang and Frangopol 2020a)

$$R_K^{net} = \sum_{i=1}^{N_{sc}} C_i^{net} \cdot p_{f,i}(t) \tag{7.24}$$

where N_{sc} is the number of failure scenarios of bridges, C_i^{net} is the monetary loss associated with the ith failure scenario, and $p_{f,i}(t)$ is the occurrence probability of the ith failure scenario at time t. The risk assessment using Eq. (7.24) should consider all possible scenarios associated with bridge failure. Since the risk assessment may require very high computation cost to address all possible scenarios, the most critical scenarios have to be selected appropriately (Saydam et al. 2013a), or the approximate assessment of the network-level risk can be used considering only the upper and lower bounds of the network-level risk (Yang and Frangopol 2020a). The monetary loss C_i^{net} consists of the direct and indirect losses. The details for estimating the direct and indirect losses of bridge network can be found in Saydam et al. (2013a), Banerjee et al. (2019), Fiorillo and Ghosn (2019), and Yang and Frangopol (2018b, 2020a).

7.3   PROBABILISTIC LIFE-CYCLE BRIDGE NETWORK MANAGEMENT

The main goals of life-cycle bridge management at a network-level are related to improving efficiency and effectiveness of life-cycle management of individual bridges under limited financial resources (Augusti et al. 1998; Liu and Frangopol 2006b; Frangopol and Bocchini 2012). The life-cycle bridge network management requires an appropriate consideration of the time-dependent network performance analysis and effects of maintenance interventions on bridge network performance and cost under uncertainty.

7.3.1   Probabilistic Time-Dependent Bridge Network Performance Analysis

The formulation of bridge network performance is based on the performance of the individual bridges in the network over time. Therefore, the prediction of time-dependent bridge network performance requires predicting the performance of individual bridges. The time-dependent performance of individual bridges can be predicted considering the deterioration mechanisms, live load effects, and environmental conditions described in Section 2.3.

The bridge network connectivity reliability p_s^{net} of Eq. (7.1) is formulated based on the reliabilities of individual bridges $p_{s,i}$. An increase in $p_{s,i}$ leads to an increase in p_s^{net}. Figure 7.3 shows both the reliability indices of individual bridges and bridge network without maintenance over time. Figure 7.3(a) shows an existing bridge network consisting of thirteen individual bridges in Colorado. The time-dependent reliability profiles of the thirteen individual bridges in Figure 7.3(b) were provided and analyzed by Akgül (2002) and Akgül and Frangopol (2004d). The series-parallel model is used to represent the bridge network, which consists of four parallel paths, and each path contains multiple individual bridges and a bridge group connected in series, as shown Figure 7.3(c). Using Eq. (7.1), the reliability index profiles of four paths and bridge network over the time horizon of 30 years can be obtained as shown in Figure 7.3(d). Furthermore, prediction of the time-dependent connectivity-based probability of failure $p_{f,CL}^{net}$ (or reliability index β_{CL}^{net}) [see Eq. (7.6)], and bridge network redundancy index RI_{net} [see Eq. (7.7)] are also based on the reliability indices of individual bridges over time.

In general, the reliability assessment and prediction of individual bridges in a network can be accomplished with extensive information related to deterioration mechanisms, live loading effects, and environmental conditions over time. When the information is not enough to predict the reliabilities of all the individual bridges in a network, the bridge life-cycle reliability profile model can be used (Frangopol et al. 2001; Kong and Frangopol 2004a). Bocchini and Frangopol (2011b) suggest several time-dependent reliability index profiles of individual bridges without maintenance. The bilinear, quadratic, square root, and exponential models for time-dependent reliability index are provided in Eqs. (7.25), (7.26), (7.27) and (7.28), respectively.

Bilinear $\qquad \beta(t) = \beta_{in}$ for $t < t_{ini}$ (7.25a)

$\qquad\qquad\qquad \beta(t) = \beta_{in} - r_\beta^B \left(t - t_{ini}\right)$ for $t \geq t_{ini}$ (7.25b)

Quadratic $\qquad \beta(t) = \beta_{in}$ for $t < t_{ini}$ (7.26a)

$\qquad\qquad\qquad \beta(t) = \beta_{in} - r_\beta^Q \left(t - t_{ini}\right)^2$ for $t \geq t_{ini}$ (7.26b)

Square root $\qquad \beta(t) = \beta_{in}$ for $t < t_{ini}$ (7.27a)

$\qquad\qquad\qquad \beta(t) = \beta_{in} - r_\beta^S \sqrt{t - t_{ini}}$ for $t \geq t_{ini}$ (7.27b)

Exponential $\qquad \beta(t) = \beta_{in}$ for $t < t_{ini}$ (7.28a)

$$\beta(t) = \left(\beta_{in} - A\right) \cdot \exp\left[-\left(\frac{t - t_{ini}}{B}\right)^2\right]$$
$$\qquad\qquad\qquad\qquad\qquad\qquad\qquad \text{for } t \geq t_{ini} \quad (7.28b)$$
$$- r_\beta^E \left(t - t_{ini}\right) + A$$

where t_{ini} is the deterioration initiation time, β_{in} is the initial reliability index, r_β^B, r_β^Q, r_β^S and r_β^E are the deterioration rates for bilinear, quadratic, square root and exponential models, respectively, A is the degradation parameter, and B is the shape parameter. Figure 7.4 shows the reliability index profiles of individual bridges based on the bilinear, quadratic, square root and exponential models.

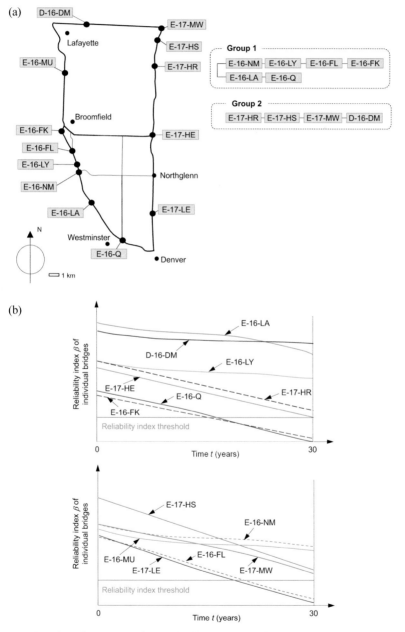

Figure 7.3   Time-dependent bridge reliability index profiles: (a) existing bridge network in Colorado; (b) reliability index profiles of individual bridges; (*Contd.*) (adapted from Liu and Frangopol 2005c, 2006b).

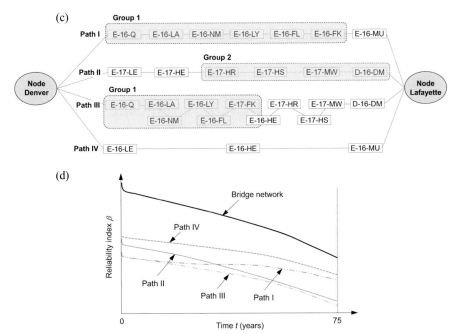

Figure 7.3 Contd. Time-dependent bridge reliability index profiles: (c) series-parallel modeling; (d) reliability index profiles of paths of series-parallel model and bridge network (adapted from Liu and Frangopol 2005c, 2006b).

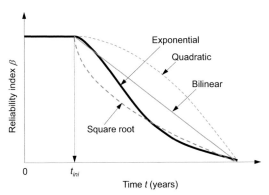

Figure 7.4 Reliability index profiles of individual bridges based on the bilinear, quadratic, square root and exponential models (adapted from Bocchini and Frangopol 2011b).

The traffic flow, time and cost of the bridge network depend on the performance of individual bridges. The connectivity level [see Eq. (7.3)], and the associated probability of failure and reliability index [see Eq. (7.6)] are affected by the annual growth rates of the communities involved in the bridge network (Bocchini and Frangopol 2013). Accordingly, the connectivity-based bridge network performance and travel flow-based bridge network performance can be predicted over time.

Also, the time-dependent cost-based and risk-based bridge network performance can be computed considering the effects of time-dependent performance of an individual bridge or a group of bridges on the maintenance cost, user cost, and direct and indirect monetary loss.

7.3.2   Life-Cycle Bridge Network Performance with Maintenances

The preventive and essential maintenance applications to individual bridges can lead to the delay and reduction of the bridge deterioration and improvement of the bridge performance, respectively (FHWA 2015; Okasha and Frangopol 2010b). Since the performance of an individual bridge affects the bridge network performance, the maintenance for individual bridges can result in the delay and reduction of bridge network deterioration and improvement of bridge network performance (Liu and Frangopol 2006b). The reliability importance of an individual bridge can be used to estimate the sensitivity of the bridge network reliability to the change in the individual bridge system reliability by applying the maintenance, and to prioritize bridge maintenance actions at a network-level (Liu and Frangopol 2005c). The reliability importance factors of an individual bridge can be computed using Eqs. (1.54) to (1.58). Figure 7.5 shows the time-dependent normalized reliability importance factors of six individual bridges in the network of Figure 7.3(a). These normalized reliability importance factors are computed based on the time-dependent reliability indices provided in Figure 7.3(b).

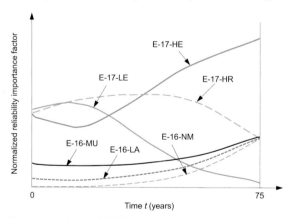

Figure 7.5   Time-dependent reliability importance of bridge network in Figure 7.3 (adapted from Liu and Frangopol 2005c).

Using the probabilistic maintenance modelings based on performance profiles, lifetime functions, damage propagation prediction, and event tree (see Section 4.3), the effect of maintenance on an individual bridge performance can be estimated, and finally, the probabilistic bridge network performance can be formulated. For example, when maintenance interventions including PM and EM are applied to three individual bridges (i.e., E-16-Q, E-16-FL and E-16-FK) in group 1 and

PM is applied to two individual bridges (i.e., E-17-HS and E-17-HR) in group 2, the associated time-dependent reliability index profiles of groups 1 and 2 are predicted, as shown in Figures 7.6(a) and 7.6(b), respectively. By taking into account the maintenance effects on the reliability indices of all the 13 individual bridges in the network of Figure 7.3(a), the bridge network reliability index can be obtained, as shown in Figure 7.6(c). The effects of maintenance interventions applied to individual bridges on bridge network performance depend on types of maintenance, maintenance application times, bridge network modeling, correlation among the individual bridges, and importance of individual bridges.

7.3.3 Probabilistic Life-Cycle Bridge Network Maintenance Management

The life-cycle bridge network performance analysis is essential for bridge owners and management agencies to efficiently manage multiple bridges in a network (Bocchini and Frangopol 2011b; Hu et al. 2015). As shown in Figure 7.7, the life-cycle performance of individual bridges in a network can be managed during a given time period by (a) allocating the financial resources for maintenance management to each individual bridge; (b) determining the expected performance to maintain each individual bridge; (c) providing the upper bound for financial resources and lower bound for the expected performance to maintain each individual bridge; and (d) optimizing the maintenance planning for all bridges in a network.

Under the allocated financial resources to each individual bridge, the bridge maintenance planning can be optimized individually by maximizing the expected bridge performance. The associated optimization process can be based on the performance-based objectives, as indicated in Section 5.3.1. When the expected performance is predefined for each individual bridge, the optimum maintenance planning can be based on minimizing the cost-based objectives presented in Section 5.3.2. If the upper bound of financial resources and the lower bound of expected performance are given for each individual bridge, the multi-objective optimization process can be applied by using the cost-based and performance-based objectives simultaneously. The optimization process for each individual bridge is required separately for bridge network management. The associated applications can be found in Chapter 6. It should be noted that the appropriate decision making process is necessary to allocate financial resources to each individual bridge, and to determine the expected performance and the upper bound for financial resources and lower bound for the expected performance for each individual bridge in a network.

Furthermore, in order to optimize maintenance planning for all the bridges in a network simultaneously, formulations of the objectives associated with bridge network performance are required. The bridge network maintenance management can provide the optimum times and types of inspection and maintenance for all the bridges in a network. This bridge network maintenance management can be optimized through a single-time optimization process.

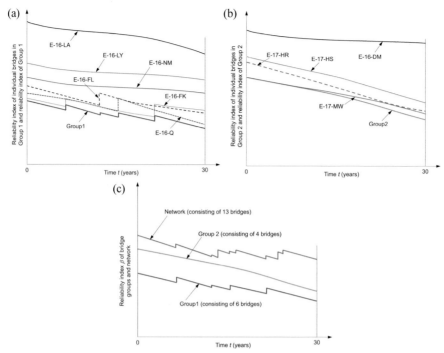

Figure 7.6 Time-dependent reliability index profiles with maintenance of several bridges: (a) individual bridges in group 1 and bridge group 1; (b) individual bridges in group 2 and bridge group 2; and (c) bridge groups and network in Figure 7.3 (adapted from Liu and Frangopol 2005a).

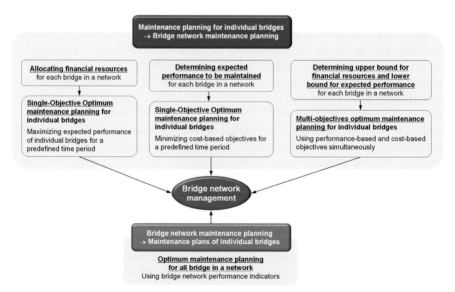

Figure 7.7 Optimum maintenance planning of all bridges in a network for a predefined time period.

7.4   OPTIMUM BRIDGE NETWORK MAINTENANCE MANAGEMENT

The bridge network maintenance management can be optimized by considering various objectives based on the bridge network performance presented in Section 7.2. The general formulation of the bridge network maintenance management is

$$
\text{Find} \qquad \mathbf{X} =
\begin{bmatrix}
x_{11} & x_{12} & \cdots & x_{1n_d} \\
x_{21} & x_{22} & \cdots & x_{2n_d} \\
\vdots & \vdots & \ddots & \vdots \\
x_{n_b 1} & x_{n_b 2} & \cdots & x_{n_b n_d}
\end{bmatrix}
\tag{7.29a}
$$

which minimizes $\quad \mathbf{\Omega}^{(-)} = \{ f_{1(-)}(\mathbf{X}), f_{2(-)}(\mathbf{X}), \ldots, f_{n(-)}(\mathbf{X}) \}$

and/or maximizes $\quad \mathbf{\Omega}^{(+)} = \{ f_{1(+)}(\mathbf{X}), f_{2(+)}(\mathbf{X}), \ldots, f_{n(+)}(\mathbf{X}) \}$ $\tag{7.29b}$

such that $\qquad O_i^- \le O_i \le O_i^+ \text{ and } x_{ij}^- \le x_{ij} \le x_{ij}^+$ $\tag{7.29c}$

where $\mathbf{X}$ is the design variable matrix, n_d is the number of design variables, n_b is the number of individual bridges in a network, and $\mathbf{\Omega}^{(-)}$ and $\mathbf{\Omega}^{(+)}$ are the objective sets consisting of objectives to be minimized and to be maximized, respectively. O_i^- and O_i^+ are the lower and upper bounds of the ith objective, respectively. x_{ij}^- and x_{ij}^+ are the lower and upper bounds of the design variable x_{ij}, respectively.

The design variables $\mathbf{X}$ for optimum bridge network maintenance management include the optimum times and types of inspection and maintenance for all the individual bridges in a network (Liu and Frangopol 2005a, 2006b; Bocchini and Frangopol 2011b; Hu and Madanat 2015). The bridge network performance indicators presented in Section 7.2 and Figure 7.1 can be used for defining $\mathbf{\Omega}^{(-)}$ and $\mathbf{\Omega}^{(+)}$. The objective set $\mathbf{\Omega}^{(-)}$ includes minimizing the cost-based, travel flow-based network disconnectedness, and risk-based bridge network performance indicators in Figure 7.1. The objective set $\mathbf{\Omega}^{(+)}$ consists of maximizing the connectivity-based network performance (see Section 7.2.1 and Figure 7.1). In order to find the Pareto front of the multi-objective optimization problem and select the most appropriate solution for bridge network maintenance management, the multi-objective optimization methods and multi-attribute decision making processes should be used as described in Section 5.4.

The optimum bridge network management based on multiple objectives has been investigated over the last few decades. Liu and Frangopol (2005a) addresses the network-level bridge maintenance planning considering two conflicting objectives simultaneously (i.e., maximizing the minimum reliability index of the bridge network connectivity between the origin and the destination locations and minimizing the expected total maintenance cost over a predefined time period of 30 years). These two objectives are formulated considering four maintenance types (i.e., resin injection, slab thickness increasing, steel plate attaching and total replacement) for enhancing performance of the deteriorating bridge deck at discrete years over the time horizon. An existing 13-bridge network located in Colorado (see Figure 7.3(a)) is used to illustrate the optimum bridge maintenance planning. The Pareto front is shown in Figure 7.8. This set indicates the types and times of maintenance applications of each of the thirteen individual bridges in the network. When the representative solutions A, B and C in Figure 7.8 are considered, the associated profiles of bridge network connectivity reliability index β^{net} and expected total maintenance cost of bridge network C_{ma}^{net} over the time horizon of 30 years can be obtained as shown in Figure 7.9. These profiles lead to maximizing β_{min}^{net} and minimizing C_{ma}^{net} over the assumed time horizon.

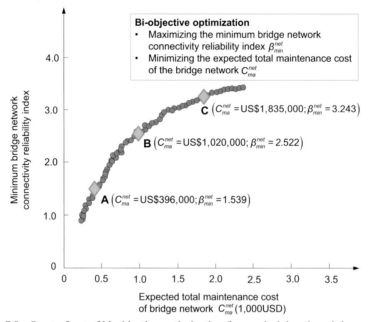

Figure 7.8   Pareto front of bi-objective optimization for maximizing the minimum bridge network connectivity reliability index and minimizing the expected total maintenance cost over the time period of 30 years (adapted from Liu and Frangopol 2005a).

For example, the optimum representative solution A in Figure 7.8 will indicate the following actions considering the time horizon of 30 years: (a) slab trickiness increasing for bridges E-17-LE, E-17-HE, E-17-HS, and E-16-MU at 19, 26, 27, and 29 years, respectively, and (b) steel plate attaching for bridge E-16-Q at 25 years. As a result, the solution A will be associated with a total maintenance cost C_{ma}^{net} of US$396,000 and a minimum value of network connectivity reliability index β_{min}^{net} of 1.539 as shown in Figures 7.8 and 7.9. The solution B in Figure 7.8 requires more maintenance actions such as (a) slab trickiness increasing for bridges E-17-LE at 14 and 19 years, E-17-HE at 15 and 26 years, E-16-FK at 24 years, and E-17-HS at 27 years, and (b) resin injection for bridges E-17-LE at 4 and 27 years, and E-17-HR at 1 and 16 years. Thus, the solution B results in larger C_{ma}^{net} and higher β_{min}^{net} than solution A. The maintenance interventions associated with the solution C will result in the reliability index profiles of individual bridges, bridge groups 1 and 2, and bridge network shown in Figure 7.6. The corresponding profiles of C_{ma}^{net} and β^{net} are presented in Figure 7.9.

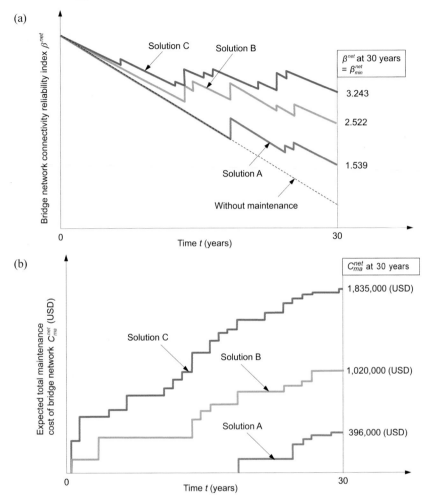

Figure 7.9 Profiles associated with solutions A, B and C in Figure 7.8, which are selected from the Pareto front of bi-objective optimization for maximizing the minimum bridge network connectivity reliability index and minimizing the expected total maintenance cost: (a) profiles of the bridge network connectivity reliability index over the time period of 30 years; and (b) profiles of the expected total maintenance cost of the bridge network (adapted from Liu and Frangopol 2005a).

Liu and Frangopol (2005a) concluded that (a) most of the Pareto solutions of the bi-objective optimization are related to the applications of resin injection and slab thickness increasing rather than steel plate attaching and total replacement, (b) applications of PM to individual bridges in a network generally leads to more cost-effective solutions than EM applications, and (c) higher reliability threshold of individual bridges may lead to EM applications and larger maintenance cost.

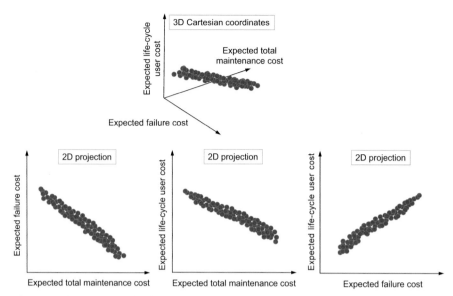

Figure 7.10 Pareto optimal solutions of tri-objective optimization for bridge network management considering minimizing the expected total maintenance cost, expected failure cost, and expected life-cycle user cost (adapted from Liu and Frangopol 2006b).

Based on the approach developed in Liu and Frangopol (2005a), the cost-based optimum bridge network maintenance management of bridge network in Figure 7.3(a) was investigated in Liu and Frangopol (2006b), where the three objectives of minimizing the expected total maintenance cost, expected failure cost, and expected life-cycle user cost are considered simultaneously. The reliability-based bridge network maintenance management was formulated as a constrained combinatorial multi-objective problem and was solved by using a genetic algorithm. The proposed framework integrates time-dependent structural reliability assessment, highway bridge network analysis, life-cycle cost calculation, and evolutionary computation. The tri-objective optimization problem results in the Pareto front shown in Figure 7.10, where the Pareto fronts in 3D Cartesian coordinates and its 2D projections are presented. Their study concluded that bridge network maintenance management is more cost-effective than individual bridge maintenance management. This procedure can cost-effectively prioritize maintenance budgets to bridges that do not meet the prescribed target performance level and/or that are most important for enhancing the network functionality over a specified time horizon.

Frangopol and Liu (2007a) presented a multi-objective optimization approach of bridge network maintenance planning by using stochastic dynamic programming. This approach consists of two phases. The first phase identifies the optimal maintenance plans for all individual bridges by minimizing the expected maintenance cost, satisfying the requirements associated with both condition and safety for a targeted time horizon. As a result, the probability mass functions of the application times for the four types of maintenance actions (i.e., minor concrete repair, silane treatment, cathodic protection, and rebuild) are

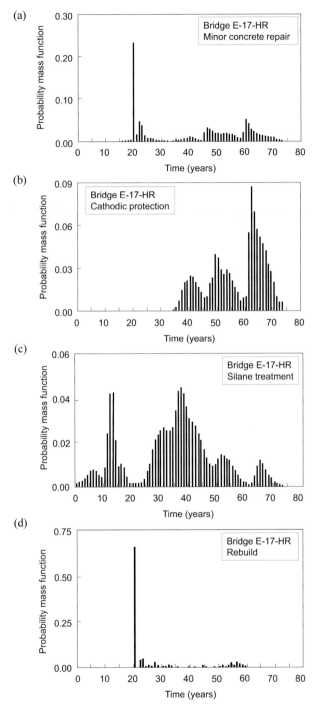

Figure 7.11   Probability mass functions of the application times for the maintenance actions: (a) minor concrete repair; (b) cathodic protection; (c) silane treatment; and (d) rebuild (adapted from Frangopol and Liu 2007a).

obtained for each individual bridge. As an example, the probability mass functions associated with each of the four maintenance types of the bridge E-17-HR in Figure 7.3(a) are illustrated in Figure 7.11. In the second phase, the optimization consists in maximizing the sum of the weighted probabilities that each of the four maintenance actions may be applied to each individual bridge in the highway network at a certain year under the constraint of a predefined yearly maintenance budget. Therefore, the annual maintenance costs for individual bridges in a network are allocated, and the most efficient combination of the maintenances for all individual bridges is obtained at a certain year.

The optimum maintenance planning at a network-level using the supply and demand approach was presented by Orcesi and Cremona (2011). The supply and demand are related to the maintenance cost and user cost, respectively. The proposed approach requires a two-step single-objective optimization. The objective is formulated as the sum of maintenance and user costs. The optimum times and types for maintenance interventions are determined from the first and second steps of optimization, respectively. Furthermore, the two objectives of minimizing the expected user cost and maintenance cost are considered for the bi-objective optimization. As a result, the Pareto front representing the times and types of maintenance is obtained.

The generalized approach for bridge network performance analysis was proposed by Bocchini and Frangopol (2011c). This approach combines three important features. The first one is the ability to compute the time-dependent reliability of the individual bridges. The second feature is the ability to deal with generalized network layouts without the need of simplified series-parallel models. The third feature is the use of a correlation among the states of individual bridges in a network. Their study concludes that the proposed approach improves the accuracy of the life-cycle bridge network reliability by considering the effect of correlation among the states of individual bridges, and leads to accurate and reliable bridge network maintenance planning.

Frangopol and Bocchini (2012) provided an extensive review on bridge network performance analysis, reliability assessment, maintenance management and optimization by dealing with spatially distributed bridges at a large scale. The modelling of extreme events and the interaction between the individual bridge and the overall network are reviewed. The large scale of bridge networks in the time domain is discussed. The previous investigations on bridge network performance assessment, and their applications are summarized. Finally, future challenges and developments, which are related to the computational efficiency, effective use of structural health monitoring data, network management with different types of systems (e.g., bridges, culverts, tunnels and pavement), and multi-disciplinary approach, are presented.

Hu and Madanat (2015) proposed an approach to ensure an adequate level of bridge network reliability at the lowest possible life-cycle maintenance cost. The presented approach, applicable to network of moderate size, requires two steps: (a) the first step results in a range of reliability levels for individual bridges in a network, and (b) in the second step, the optimal reliability levels of individual bridges are determined by minimizing the total maintenance cost for maintaining a predefined network reliability level. Their study shows that the maintenance cost for a bridge network increases with an increase in the reliability level of individual bridges. The network connectivity problem can be solved by using the proposed approach, which makes it possible to use standard optimization tools. Furthermore, Hu et al. (2015) developed a simplified framework to optimize the life-cycle budgets for maintenance interventions on large scale bridge networks. The optimum life-cycle maintenance interventions are associated with the minimum expected increase in user cost due to bridge failure.

Bai et al. (2013) proposed a method to evaluate the network performance of each candidate maintenance intervention for individual bridges and to identify the optimal maintenance intervention associated with the maximum bridge network utility. The bridge network utility consists of four average condition ratings of individual bridges (i.e., average superstructure condition rating, average substructure condition rating, average deck condition rating, and average wearing surface condition rating). The maximum bridge network utility results in the optimum average condition ratings of individual bridges for a given financial budget level. Their study shows that the bridge network-based maintenance intervention is more effective than individual-based maintenance intervention.

A framework to determine preferred maintenance interventions for bridge networks, where the optimization uses a single objective of minimizing the total travel time, was presented by Zhang and Wang (2017). This framework adopts modern network analysis methods, structural reliability principles and meta-heuristic optimization algorithms to formulate and solve the optimization problem. The total travel time is estimated considering the bridge capacity ratings, condition ratings, traffic demands, and locations of the individual bridges in a network. From the optimization process, individual bridges for renewal or replacement are determined for a given financial budget.

Kim et al. (2020) proposed a probabilistic approach to determine the target reliability level of individual bridges for service life management at the bridge network level. The target reliability levels of individual bridges are determined through a multi-objective optimization process based on four objectives: maximizing the reliability, minimizing the expected maintenance cost, minimizing the expected user costs, and maximizing the redundancy of the bridge network. The design variables of this optimization problem are the target reliability indices of the individual bridges in the network. The illustrative example presented is an existing bridge network consisting of 57 individual bridges. Their study shows that the optimum target reliability indices for individual bridges have to be different in order to obtain a larger reliability, lower maintenance cost, lower user cost, and larger redundancy of the bridge network than in the case where the target reliability of all bridges is the same.

7.5   CONCLUDING REMARKS

This chapter deals with probabilistic life-cycle bridge network management. The bridge network management requires assessment and prediction of bridge network performance under uncertainty. The bridge network performance indicators are: (a) connectivity-based; (b) travel flow-based; (c) cost-based; and (d) risk-based. The general concepts of bridge network performance with maintenance interventions and life-cycle bridge network maintenance management are presented. The formulation of the optimum life-cycle bridge network maintenance management based on multiple bridge network performance is described. The associated applications are summarized. The probabilistic life-cycle bridge network management presented in this chapter considers the performance deterioration of individual bridges under normal loading effects and environmental conditions. The bridge network management under extreme events (e.g., earthquakes, hurricanes, and floods) is provided in Chapter 8.

Chapter 8

Resilience and Sustainability of Bridges and Bridge Networks

ADD_{ij}	= average daily traffic detoured at the ith link in damage state j
ADR_{ij}	= average daily traffic remaining at the ith link in damage state j
ADT_{ij}	= average daily traffic at the ith link in damage state j
$C_{con|ds,k}$	= consequence conditioned by the kth damage state
$C_{etr,ij}$	= monetary loss associated with the jth damage state of the ith link
C_R	= total expected retrofit cost
$C_{SU}^{IN}(t)$	= monetary loss in terms of sustainability
CB	= cost-benefit indicator with retrofit
$E[C_{EC}(t)]$	= expected total economic monetary loss at time t
$E[C_{EC,fi}(t)]$	= expected life loss of fatalities at time t
$E[C_{EC,rp}(t)]$	= expected repair and construction cost for damaged or failed bridges at time t
$E[C_{EN}(t)]$	= expected total environmental monetary loss at time t
$E[C_{SC}(t)]$	= expected total social monetary loss at time t
$E[C_{SU}^{NET}(t)]$	= total monetary loss in terms of sustainability at time t
$E[t_{etr}(t)]$	= expected extra travel time at time t
$E[W_{EN,tr}(t)]$	= expected environmental metric related to the traffic detour after an earthquake at time t
$E[W_{EN,ew}(t)]$	= expected environmental metric related to energy waste at time t
$F(t|e)$	= system state at time t under an event e_i
$P_{bds,ij|GI}(t)$	= probability of the ith bridge being in the jth damage state after an earthquake with a certain ground motion intensity at time t
$P_{ds,k|ls,j}(t)$	= conditional occurrence probability of the kth damage state given the jth structural limit state at time t
$P_{lds,ij|GI}(t)$	= probability of the ith link being in the jth damage state after an earthquake with a certain ground motion intensity at time t
$P_{ls,j|ex,j}(t)$	= occurrence probability of the jth structural limit state conditioned by the ith extreme event at time t

NOTATIONS (continued)	
$P_{ex,i}(t)$	= occurrence probability of the ith extreme event at time t
L_{dmg}	= damage level
$m[G_{i,m}(t)]$	= median of ground motion intensity for a bridge in the ith damage state at time t
N_{bdg}	= number of bridges in a network
N_{ex}	= number of extreme events
N_l	= number of links in a bridge network
N_{ls}	= number of structural limit states
N_{ds}	= number of damage states
Q_{loss}	= functionality loss
Q_{tg}	= target functionality
$Q(t)$	= time-dependent functionality of a bridge
r_{AADT}	= ratio of average daily truck traffic to average daily traffic
$r_{ma,i}$	= enhancement ratio of ground motion for a bridge in the ith damage state after maintenance
R_{lev}	= retrofit level
RB	= robustness
RD	= rapidity of functional recovery
RD_0	= rapidity of functional recovery defined as an angle
RS	= resilience
RS_{loss}	= loss of resilience
RS_{loss}^{life}	= lifetime resilience loss
t_0	= time of hazard occurrence
$t_{d,ij}$	= down time associated with the jth damage state of the ith link
$t_{etr,ij}$	= extra travel time of the ith link for the jth damage state
t_{id}	= idle time period
t_{rec}	= functional recovery time period
U_B	= benefit utility due to the retrofit actions
U_C	= utility associated with seismic retrofit cost
$U_{S,0}$	= multi-attribute utilities without retrofit
$U_{S,R}$	= multi-attribute utilities with retrofit
γ	= parameter to represent the risk attitude of the decision maker

ABSTRACT

Chapter 8 provides the general concepts to estimate resilience and sustainability for the management of individual bridges and bridge networks under extreme events. The effects of bridge performance deterioration and hazard-induced damage, and risk mitigation on resilience are investigated. The representative functionality recovery models to compute resilience are provided. Applications of resilience assessment for bridge and bridge network management are provided. Furthermore, this chapter presents the sustainability assessment and optimum bridge network management based on sustainability. The sustainability assessment considers the economic, social, and environmental consequences of bridge failures in a holistic manner. Significant applications associated with sustainability assessment and optimum bridge and bridge network management are provided.

8.1 INTRODUCTION

Bridges are among the most vulnerable components of highway networks under extreme events such as natural and human-made hazards including earthquakes, tsunamis, floods, hurricanes, and terrorist attacks (Liu and Frangopol 2006b; Saydam et al. 2013a; Stewart and Mueller 2014; Li et al. 2020; Yang and Frangopol 2020b; Ishibashi et al. 2021). As the performance of bridges and bridge networks deteriorates over time, they become more vulnerable to hazards and their associated risks increase (Choe et al. 2009; Akiyama et al. 2011, 2013; Dong et al. 2014a, 2014b). In order to manage the life-cycle risk of bridges and bridge networks under extreme events efficiently, it is necessary to minimize direct and indirect losses by improving their resistance and robustness to extreme events and maximize their efficiency of functional recovery (Frangopol and Bocchini 2012; Decò et al. 2013; Alipour and Shafei 2016). Since the last decade, resilience and sustainability have been considered as useful performance indicators for bridge and bridge network management under extreme events. Accordingly, several approaches to assess resilience and sustainability have been developed and applied to optimum bridge network management planning under single and multiple hazards (Rokneddin et al. 2013; Bocchini et al. 2014; Dong and Frangopol 2019; Akiyama et al. 2020).

In this chapter, resilience and sustainability for management of individual bridges and bridge networks under extreme events are addressed. The general concept of resilience of bridges and bridge networks, including the effects of bridge performance deterioration and hazard-induced damage and risk mitigation, is presented. The representative functionality recovery models to compute resilience are provided. In order to consider the economic, social, and environmental consequences of bridge failures holistically, the concept of sustainability and its assessment are provided. Representative applications of resilience and sustainability assessment for bridges and bridge networks are summarized.

8.2 RESILIENCE OF BRIDGES AND BRIDGE NETWORKS

The concept of resilience has continuously evolved over the last few decades with increasing demand to adopt resilience in economics, environmental science, social science, and engineering. Resilience generally indicates the ability of a system to recover efficiently from a severe shock and to maintain functionality when shocked (Holling 1973; Perrings 2006; Rose 2004). Since the last few decades, resilience has been used to represent the performance of various civil infrastructure systems such as water supply systems, power transmission systems, and transportation systems under extreme events such as hurricanes, earthquakes, or large-scale terrorist attacks (Banerjee et al. 2019).

8.2.1   General Concept of Resilience for Bridge Management

Resilience for bridge management is used to quantify the promptness of the restoration of individual bridges and bridge networks after extreme events (Bocchini et al. 2012). Bruneau et al. (2003) describe a resilient system as a system associated with (a) reduced probability failure; (b) reduced damage and negative economic and social consequences; and (c) reduced time to recovery. Since resilience covers social and technical aspects too broadly, there is no single generally accepted definition (Decò et al. 2013). For example, the resilience RS after a single extreme event can be defined as (Bruneau et al. 2003; Bruneau and Reinhorn 2007; Cimellaro et al. 2010; Bocchini and Frangopol 2012a)

$$RS = \int_{t_0}^{t_0+t_h} Q(t)\,dt \tag{8.1}$$

where t_0 is the time of extreme event occurrence, t_h is the investigated time horizon, and $Q(t)$ is the time-variant functionality of a bridge. The resilience RS of Eq. (8.1) corresponds to the area under the functionality curve as shown in Figure 1.15. If there is no reduction of functionality after a single extreme event, the resilience RS from t_0 the time $t_0 + t_h$ is equal to $Q(t_0) \times t_h$. According to Bruneau et al. (2003), the loss of resilience RS_{loss} due to an extreme event is defined by

$$RS_{loss} = \int_{t_0}^{t_0+t_h} \{1 - Q(t)\}\,dt \tag{8.2}$$

When there is no reduction of functionality after a single extreme event, the loss of resilience RS_{loss} becomes zero. Assuming that the functionality is restored to the initial value after rehabilitation (i.e., $Q(t_0 + t_{rec}) = 1.0$), the resilience of Eq. (8.1) can be normalized as (Frangopol and Bocchini 2011; Bocchini and Frangopol 2012b)

$$RS = \frac{\int_{t_0}^{t_0+t_h} Q(t)\,dt}{t_h} \tag{8.3}$$

When the functionality is restored to a predefined target level of functionality Q_{tg}, which is at most one, the resilience can be expressed as

$$RS = \frac{\int_{t_0}^{t_0+t_h} Q(t)\,dt}{t_h \cdot Q_{tg}} \qquad \text{for} \qquad Q_{tg} \le 1.0 \tag{8.4}$$

If the target level of functionality Q_{tg} is defined as the functionality $Q(t_0')$ just before the time of occurrence of the extreme event, the resilience can be estimated as

$$RS = \frac{\int_{t_0}^{t_0+t_h} Q(t)\,dt}{t_h \cdot Q(t_0')} \qquad \text{for} \qquad Q_{tg} = Q(t_0') \leq 1.0 \tag{8.5}$$

Moreover, Henry and Ramirez-Marquez (2012) define resilience as

$$RS\left(t_r \middle| e_i\right) = \frac{F\left(t_r \middle| e_i\right) - F\left(t_d \middle| e_i\right)}{F_{in} - F\left(t_d \middle| e_i\right)} \tag{8.6}$$

where $RS(t_r|e_i)$ indicates the resilience at time t_r under a disruptive event e_i, F_{in} is the initial state of the system, and $F(t_r|e_i)$ and $F(t_d|e_i)$ are the system states at times t_r and t_d under a disruptive event e_i, respectively. The time t_r is between the time of disruption t_d and the investigated time horizon t_h [i.e., $t_r \in (t_d, t_h)$]. If the system does not recover from a disruptive state [i.e., $F(t_r|e_i) = F(t_d|e_i)$], the resilience $RS(t_r|e_i)$ becomes zero. If the system recovers from a disruptive state to the original stable state at time t_r [i.e., $F_{in} = F(t_r|e_i)$], $RS(t_r|e_i)$ will be equal to one.

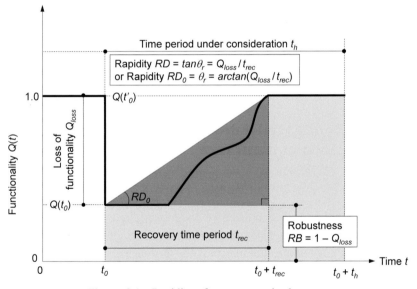

Figure 8.1 Rapidity of recovery and robustness.

The representative properties of resilience include rapidity, robustness, redundancy, and resourcefulness (Bruneau et al. 2003; Cimellaro et al. 2010). Figure 8.1 illustrates these properties of resilience.

- Rapidity indicates the capacity to achieve goals promptly to avoid future disruption. As shown in Figure 8.1, the rapidity RD is represented by an average slope of the functionality curve during the recovery time period t_{rec} as (Cimellaro et al. 2010; Decò et al. 2013)

$$RD = \frac{Q_{loss}}{t_{rec}} \tag{8.7a}$$

or

$$RD_0 = \arctan\left(\frac{Q_{loss}}{t_{rec}}\right) \tag{8.7b}$$

where Q_{loss} is the loss of functionality due to an extreme event, which is the difference between the functionalities at time t_0 and $t_0 + t_{rec}$ [i.e., $Q_{loss} = Q(t_0 + t_{rec}) - Q(t_0)$]. The reduction in recovery time period t_{rec} results in an increase of the rapidity.

- Robustness associated with resilience shows the strength of a system to withstand a given stress (Bruneau et al. 2003). As shown in Figure 8.1, the robustness

$$RB = 1 - Q_{loss} \tag{8.8}$$

can be quantified as the residual functionality after an extreme event.
- Redundancy represents the capability of a structural system to carry loads after the damage of one or more of its members and to satisfy functional requirements (Frangopol and Curley 1987; Ghosn et al. 2010).
- Resourcefulness is the capacity to identify problems, establish priorities, and mobilize resources when there are conditions threatening to disrupt the system. Resourcefulness can increase by applying additional resources (e.g., monetary, physical, technological, and informational) and human resources in the recovery process.

8.2.2   Effects of Deterioration, Damage and Risk Mitigation on Resilience

Resilience can be affected by functionality deterioration and hazard-induced damage and risk mitigation of bridges. Figure 8.2 shows the functionality profiles of bridges with and without functionality deterioration, where the functionality is recovered linearly with time to the target level Q_{tg} equal to 1.0. In this figure, it is assumed that there is no deterioration after the loss of functionality until the target functionality is reached. The functionality of a bridge decreases continuously over time, along with the bridge performance deterioration under external loadings and environmental conditions. In Figure 8.2(a), the functionality loss Q_{loss} due to an extreme event and rapidity of recovery RD_0 for bridges with and without functionality deterioration are the same. The area under the functionality curve without deterioration is larger than the area associated with deterioration. According to Eqs. (8.1), (8.3) and (8.4), the resilience of the bridge without functionality deterioration is larger than that of the deteriorating bridge. As shown in Figure 8.2(b), assuming that the functionality loss Q_{loss} and recovery time period t_{rec} are the same for the bridges with and without functionality deterioration, the resilience of the bridge without functionality deterioration is larger than that of the bridge with functionality deterioration. The same conclusion is associated with the case of bridges with the same robustness, rapidity and recovery time period, as shown in Figure 8.2(c).

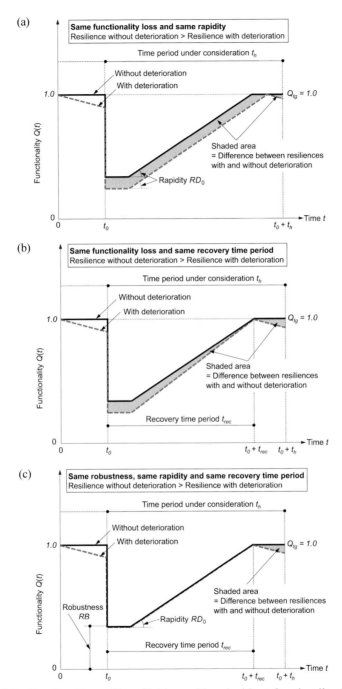

Figure 8.2 Functionality profiles of bridges with and without functionality deterioration with a fixed target recovery level: (a) same functionality loss and same rapidity; (b) same functionality loss and same recovery time period; (c) same robustness, same rapidity and same recovery time period.

The functionality profiles of bridges without and with functionality deterioration restored after an extreme event to the full functionality (i.e., $Q_{tg} = 1.0$) and to the functionality level just before the time of the occurrence of the extreme event (i.e., $Q_{tg} = Q(t_0') < 1.0$), respectively, are compared in Figure 8.3. In this figure, it is assumed that the two bridges have the same functionality loss Q_{loss}, rapidity RD_0, and recovery time period t_{rec} regardless of functionality deterioration. It is also assumed that there is no deterioration during the recovery time period. According to Eq. (8.5), the resilience of the bridge without functionality deterioration will be larger than that with functionality deterioration.

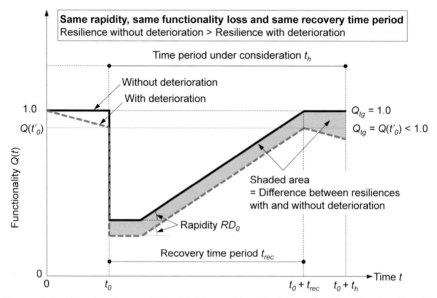

Figure 8.3 Functionality profiles of bridges with and without functionality deterioration with a target functionality level equal to the functionality just before the time of the extreme event occurrence.

The recovery patterns associated with several damage levels (i.e., no damage, slight damage, moderate damage, and extensive damage) are illustrated in Figure 8.4(a). The slight damage causes the fastest recovery. The functionality profile associated with moderate damage can be represented by a stepwise-type curve, since the partial recovery is first performed, and then there is an idle period. After the extensive damage, the recovery pattern can be associated with a positive exponential-type pattern, considering that the bridge has full functionality only after rehabilitation (Decò et al. 2013). Figure 8.4(b) shows the recovery patterns for complete damage with and without bypass. The complete damage results in no functionality. If a temporary bypass is built to carry a portion of the original traffic flow after complete damage, a stepwise-type recovery pattern can be expected. To increase the functionality of a bridge after complete damage, temporary options such as a bypass can be applied.

Risk mitigation alternatives for bridge network management are used to reduce the probability of bridge failure and functionality loss under specific extreme

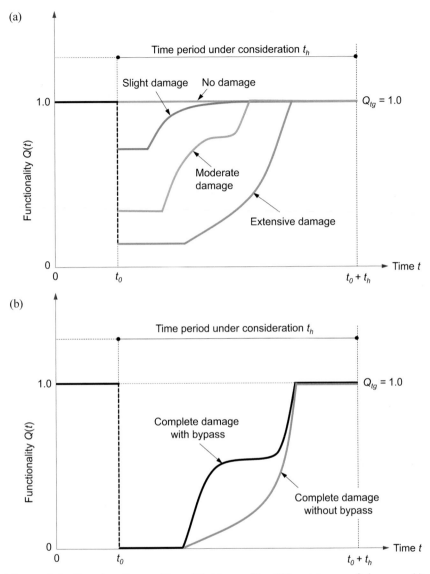

Figure 8.4 Functionality profiles of bridges with different types of damage: (a) no damage, slight damage, moderate damage, and extensive damage; (b) complete damage with and without bypass (adapted from Decò et al. 2013).

events (Zhu and Frangopol 2013a). The effect of risk mitigation on the bi-linear functionality profile is shown in Figure 8.5. In this figure, it is assumed that the functionality is recovered to the target level Q_{tg} equal to 1.0, and no deterioration is considered. When the rapidity of recovery RD_0 is fixed (i.e., the same for the bridges with and without risk mitigation), and risk mitigation alternatives result in reduction of the functionality loss Q_{loss}, the bridge with risk mitigation has a larger resilience than the bridge without risk mitigation, as shown in Figure 8.5.

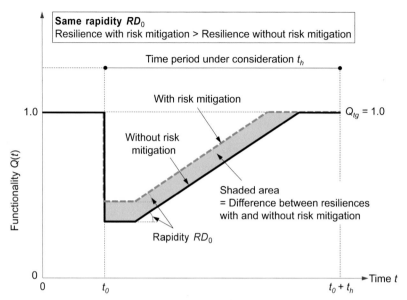

Figure 8.5   Functionality profiles of bridges with and without risk mitigation when the same rapidity of recovery is applied.

It can be concluded that the resilience of a bridge increases by delaying the deterioration initiation and reducing the functionality deterioration rate (see Figures 8.2 and 8.3), adopting the partial recovery options (e.g., temporary roads and bridges for bypass) (see Figure 8.4(b)), and using the risk mitigation to reduce the functionality loss induced by an extreme event (see Figure 8.5). Using preventive maintenance (PM), the deterioration initiation can be delayed, and the functionality deterioration rate and probability of failure for specific hazards can be reduced. The essential maintenance (EM) and replacement (RP) can improve the bridge performance and reduce the probability of failure. The details on PM, EM and RP are described in Section 4.2. Moreover, the seismic risk mitigation alternatives for bridges include elastomeric bearings, set extenders, restrainer cables, shear keys and steel jack as well as PM, EM and RP, which can reduce the probability of failure and functionality loss (Padgett et al. 2010; Zhu and Frangopol 2013a).

8.2.3   Probabilistic Resilience

The resilience of a bridge can be represented by the parameters such as robustness RB (or residual functionality), idle time period t_{id}, recovery duration t_{rec}, and target functionality Q_{tg} (Decò et al. 2013). By treating these parameters as random variables, the functionality of an individual bridge under uncertainty can be illustrated as shown in Figure 8.6. The robustness RB of a bridge requires assessment of the post-event damage. The fragility analysis provides the damage states (no damage, slight, moderate, extensive, and complete damage) caused by

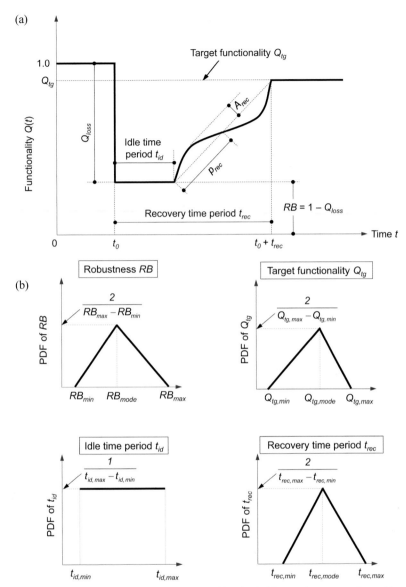

Figure 8.6 Representative functionality profile of a bridge under uncertainty: (a) profile and its parameters and random variables; (b) PDFs of random variables (adapted from Bocchini et al. 2012, and Decò et al. 2013).

various hazards and the associated occurrence probabilities. The software HAZUS (FEMA 2020) can be used to compute the potential losses from earthquakes, floods, and hurricanes. For instance, HAZUS provides the probabilistic results on the fragility analysis of each bridge for a given earthquake scenario. Using these results, the expected damage level of an individual bridge $E(d_{ind})$, taking values in the interval [0, 4], can can be computed as (Bocchini and Frangopol 2011a)

$$E(d_{ind}) = 0 \cdot P(d_{no}) + 1 \cdot P(d_{sl}) + 2 \cdot P(d_{md}) + 3 \cdot P(d_{ext}) + 4 \cdot P(d_{com}) \quad (8.9)$$

where $P(\cdot)$ is the probability of being in a specific damage state as computed by HAZUS, and d_{no}, d_{sl}, d_{md}, d_{ext}, and d_{com} indicate no damage, slight, moderate, extensive, and complete damage states, respectively.

During the idle time period t_{id}, initial plans for bridge rehabilitation can be established considering the rapidity, robustness, redundancy, and resourcefulness of a bridge network resilience. Specific activities not to improve the bridge functionality (e.g., demolition of a damaged bridge or removal of the debris) will affect the idle time t_{id}. The recovery pattern depends on the recovery duration, rapidity, functionality loss, and rehabilitation and reconstruction process. The recovery model can be formulated using several representative shapes such as stepwise, linear, positive and negative exponential, and sinusoidal curves. For example, the recovery pattern in Figure 8.6 is represented by the sinusoidal curve expressed by the two parameters p_{rec} and A_{rec}. These two parameters are associated with the position of the inflection point and the amplitude of the sinusoidal curve, respectively. Decò et al. (2013) show that the robustness RB, idle time period t_{id}, recovery time period t_{rec}, and target functionality Q_{tg} can be treated as random variables, represented by the probability density functions (PDFs) in Figure 8.6(b). The two parameters p_{rec} and A_{rec} depend on the damage types shown in Figure 8.4(a).

8.2.4   Functionality Recovery Models

Functionality recovery models have been introduced by several investigations such as Kafali and Grigoriu (2005), Padgett and DesRoches (2007), Cimellaro et al. (2010), Bocchini et al. (2012), and HAZUS (FEMA 2020). Several functionality recovery models are as follows:

(a)   Linear functionality recovery model (Bocchini et al. 2012)

$$Q(t) = RB + H\left[t - t_0 - t_{id}\right] \cdot RD \cdot \left(t - t_0 - t_{id}\right) \quad (8.10)$$

where RB is the robustness [see Eq. (8.8)], $H(\cdot)$ is the Heaviside unit step function, and RD is the rapidity [see Eq. (8.7a)].

(b)   Trigonometric functionality recovery model (Bocchini et al 2012)

$$Q(t) = RB + H\left[t - t_0 - t_{id}\right] \cdot \frac{Q_{tg} - (1 - Q_{loss})}{2} \cdot \left[1 - \cos\left(\frac{\pi\left(t - t_0 - t_{id}\right)}{t_{rec} - t_{id}}\right)\right] \quad (8.11)$$

(c)   Exponential-based functionality recovery model (Kafali and Grigoriu 2005)

$$Q(t) = RB + H[t - t_0 - t_{id}] \cdot [Q_{tg} - (1 - Q_{loss})] \cdot \left[1 - \exp\left(\frac{-k\left(t - t_0 - t_{id}\right)}{t_{rec} - t_{id}}\right)\right] \quad (8.12)$$

where k is the shape parameter. When there is no idle time period (i.e., $t_{id} = 0$), the functionality recovery models of Eq. (8.10), (8.11), and (8.12) can be simplified, respectively, as (Cimmelaro et al. 2010)

$$Q(t) = RB + RD \cdot (t - t_0) \tag{8.13a}$$

$$Q(t) = RB + \frac{Q_{tg} - (1 - Q_{loss})}{2} \cdot \left[1 - \cos\left(\frac{\pi (t - t_0)}{t_{rec}} \right) \right] \tag{8.13b}$$

$$Q(t) = RB + \left[Q_{tg} - (1 - Q_{loss}) \right] \cdot \left[1 - \exp\left(\frac{-k(t - t_0)}{t_{rec}} \right) \right] \tag{8.13c}$$

(d) Functionality recovery model based on normal cumulative distribution function (CDF) used in HAZUS (FEMA 2020)

$$Q(t) = RB + \left[Q_{tg} - (1 - Q_{loss}) \right] \cdot erf\left(\frac{t - t_0 - \lambda}{\sqrt{2}\rho} \right) \tag{8.14}$$

where $erf(\cdot)$ is the error function, and λ and ρ are the shape parameters depending on the recovery time period t_{rec}.

(e) Stepwise functionality recovery model (Padgett and DesRoches 2007)

$$Q(t) = \begin{cases} RB & t_0 \leq t < t_1 \\ Q_1 & t_1 \leq t < t_2 \\ \vdots & \vdots \\ Q_n & t_n \leq t < t_0 + t_{rec} \\ Q_{tg} & t \geq t_0 + t_{rec} \end{cases} \tag{8.15}$$

where Q_i = values of the functionality between robustness RB and target functionality Q_{tg} achieved at the time instant t_i. It should be noted that formulation of the time-dependent functionality recovery models $Q(t)$ presented in Eqs. (8.10) to (8.15) are available after an extreme event occurred at time t_0 (i.e., $t \geq t_0$). Eq. (8.15) considers that a bridge traffic flow capacity depends on the available number of lanes, and therefore, a stepwise functionality recovery model may be more appropriate than a continuous recovery model.

(f) Generalized trigonometric model

$$Q(t) = RB + H[t - t_0 - t_{id}] \cdot f_I(t - t_{id}) \tag{8.16}$$

Considering that functionality can reach its target level Q_{tg} at the end of the rehabilitation, the partial functionality recovery function $f_I(\cdot)$ in Eq. (8.16) is

$$f_I(z) = \frac{Q_{tg} - RB}{1 - RB} \cdot f_{II}(z) \tag{8.17}$$

where $f_{II}(z)$ is the adjust function to represent the location of the sinusoid flex. The detailed formulation of $f_{II}(z)$ and the effect of the parameters on the proposed trigonometric model can be found in Bocchini et al. (2012).

8.3   APPLICATIONS OF RESILIENCE FOR OPTIMUM BRIDGE NETWORK MANAGEMENT

Resilience can be used for decision making related to restoration of individual bridges and bridge networks after extreme events. Accordingly, probabilistic approaches to evaluate resilience, and to establish optimum restoration strategies of bridge networks to maximize resilience have been developed since the last decade (Frangopol and Bocchini 2012; Dong and Frangopol 2019). In this section, several significant applications for resilience assessment and bridge network management based on resilience are provided.

8.3.1   Applications of Resilience Assessment for Individual Bridges and Bridge Networks

A probabilistic approach to evaluate the seismic resilience of bridges and to assess direct and indirect costs was developed by Decò et al. (2013). This approach predefines possible restoration strategies related to four recovery velocities (i.e., do nothing, slow, average, and fast recovery) with and without bypass and four damage levels (i.e., slight, moderate, extensive, and complete damage). For each restoration strategy, a fragility analysis is performed using the computer program HAZUS, and time-dependent functionality, resilience, and rapidity are computed, considering the uncertainties associated with expected damage, restoration process, and rebuilding/rehabilitation costs. As an illustrative example, an existing bridge network located between the cities of Corona and Murrieta in California was adopted. Figure 8.7 shows the mean of resilience, rapidity, and direct and indirect costs of 12 out the 24 potential restoration strategies reported by Decò et al. (2013). Their study concluded that (a) the expected functionality recovery is highly affected by restorations associated with extensive and complete damages; (b) the restoration strategy requiring the highest cost limits the indirect consequences and results in the highest expected resilience and shortest expected full recovery time; (c) the adoption of temporary bypass in the case of complete damage can mitigate the impact of indirect cost without significant increase of resilience; and (d) the decision makers should take into account the target bridge performance level after restoration, allowable economic impact, and available financial resources for efficient post-event restoration.

A probabilistic seismic performance assessment of bridges subjected to mainshock and aftershocks of an earthquake was reported by Dong and Frangopol (2015). The proposed approach consists of seismic event identification and updating, and assessments of seismic performance, risk and resilience. The seismic event identification generates an initial seismic scenario, and updates the magnitude and source-to-site distance associated with the initial seismic scenario using Bayes' theorem with historical data and real-time information. Through the seismic performance assessment, the fragility curves of a bridge under mainshock only (MS) and mainshock and aftershocks (MSAS) are determined. From these curves, economic direct and indirect losses, risk, functionality, and resilience of a bridge are computed. The proposed approach is illustrated with a two-lane two-span RC bridge designed based on Caltrans's bridge seismic design criteria (CSDC 2004). Figure 8.8 shows the expected functionality $E(Q)$ and expected daily indirect loss $E(C_{ind})$ over time associated with MS and MSAS. As shown in Figure 8.8, aftershocks reduce the expected functionality (see Figure 8.8(a)) and increase the expected daily indirect costs (see Figure 8.8(b)) compared to the case when only the main shock is considered.

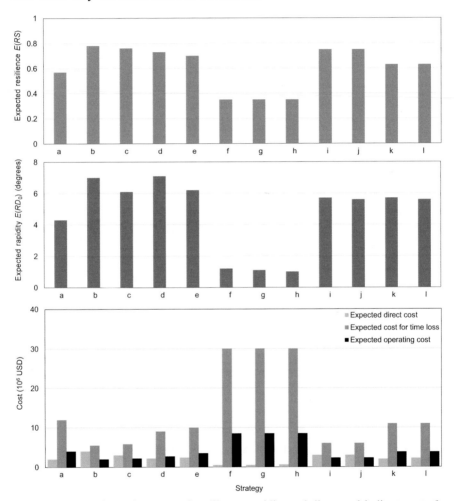

Figure 8.7   Bar charts for expected resilience, rapidity and direct and indirect costs for 12 restoration strategies (adapted from Decò et al. 2013).

Biondini et al. (2015) presented a probabilistic seismic resilience of deteriorating concrete structures, considering the interaction between environmental aggressiveness and seismic events. Their approach integrates the effects of (a) damage due to environmental aggressiveness, (b) maintenance intervention before seismic events, and (c) post-seismic event recovery actions on the time-variant functionality of a bridge. Recovery functions represented by negative-exponential, sinusoidal, and positive-exponential models are considered. As illustrative examples, a three-story concrete frame and a four-span continuous RC bridge are presented. Their study shows that (a) corrosion of reinforcement can reduce the expected functionality of a structural system; and (b) a multi-hazard life-cycle oriented approach to seismic design of resilient structure and infrastructure systems can play a significant role to limit the effects of aging and deterioration and to establish effective and rapid post-event recovery procedures.

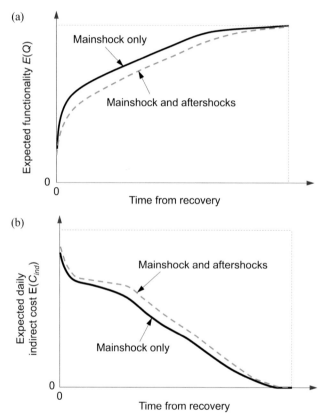

Figure 8.8   Effects of mainshock only (MS) and mainshock and aftershocks (MSAS) on: (a) expected functionality $E(Q)$ and (b) expected daily indirect loss $E(C_{ind})$ of a bridge (adapted from Dong and Frangopol 2015).

A generalized framework to assess the resilience of bridge networks under deterioration processes and seismic events was developed by Alipour and Shafei (2016). This framework is applied to a bridge network consisting of nine types of RC bridges located in Southern California. Time-dependent performance of individual bridges under corrosion is predicted using the approaches developed by Shafei et al. (2012). A set of earthquakes with different occurrence probabilities and several aging scenarios are adopted to evaluate bridge damage states, direct and indirect losses, and resilience of bridge networks. Their study highlights that extensive deterioration of the network components may dramatically increase the number of damaged bridges and indirect losses after earthquakes, and the individual bridges should be ranked based on their importance to manage the functionality of the bridge network efficiently.

8.3.2 Optimal Resilience-Based Management of Individual Bridges and Bridge Networks

A novel framework for decision-making on optimum bridge interventions after an extreme event was proposed by Bocchini and Frangopol (2012a). The proposed framework was illustrated with three bridge network examples (i.e., series, series with crossing road, parallel and series-parallel configurations). The series, parallel, and series-parallel configurations are indicated in Figure 8.9. It is assumed that the structure types and damage types of these three bridges are the same, and a bi-linear restoration model associated with damage levels is used as shown in Figure 8.10.

The damage level L_{dmg} ranges from 0 (i.e., no damage) to 4 (total bridge collapse). In their study, the resilience is defined as the integral in time of the functionality (i.e. Eq. (8.1)). The three bridges are assumed to be strongly damaged by an extreme event (i.e., $3 \le L_{dmg} \le 4$). The idle time periods t_{id} and speed of restorations θ_r (see Figure 8.10) of the three bridges in the network are the design variables of the bi-objective optimization. The objectives are maximization of the bridge network resilience and minimization of the expected total restoration cost. The Pareto fronts of the bi-objective optimization for the three bridge network examples in Figure 8.9 are presented in Figure 8.11. These Pareto fronts are grouped into (a) no bridge

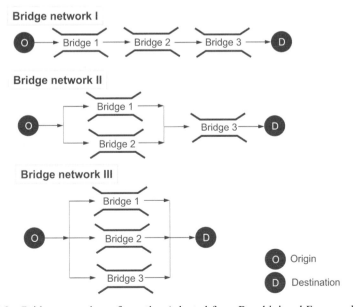

Figure 8.9 Bridge network configuration (adapted from Bocchini and Frangopol 2012a).

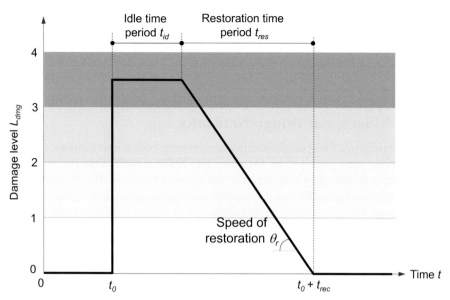

Figure 8.10   Bridge restoration model (adapted from Bocchini and Frangopol 2012a).

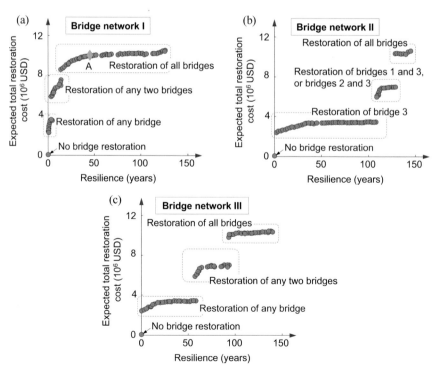

Figure 8.11   Pareto fronts for bridge network management based on resilience: (a) bridge network I; (b) bridge network II; (c) bridge network III in Figure 8.9 (adapted from Bocchini and Frangopol 2012a).

restoration, (b) one bridge restoration, (c) two bridge restorations, and (d) all bridge restorations. The Pareto optimal solutions associated with all bridge restorations require higher expected total restoration cost and result in larger resilience than other Pareto optimal solutions. Figure 8.12 shows the evolution of the restoration process during time of each of the three bridges considered and the functionality profile of the bridge network associated with Solution A, which requires all bridge restoration as shown in Figure 8.11(a). It should be noted that the functionality used in Bocchini and Frangopol (2012a) ranges from 0 (i.e., not functioning) to 100% (i.e., fully functioning). From Figure 8.12, it can be seen that the functionality of the bridge network becomes 100%, when all the restorations are completed. Resilience is defined as the normalized integral over time of the network functionality [see Eq. (8.4)].

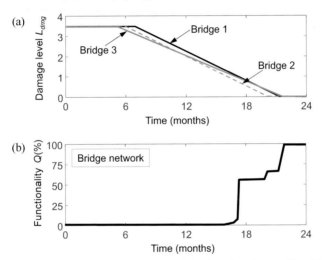

Figure 8.12 (a) Damage levels of bridges and (b) functionality profile of bridge network associated with solution A in Figure 8.11(a) (adapted from Bocchini and Frangopol 2012a).

Based on the approach developed by Bocchini and Frangopol (2012a), the tri-objective optimum restoration activities were investigated in Bocchini and Frangopol (2012b). The restoration activities are optimized to maximize the bridge network resilience, minimize the time required to reach a target functionality level, and minimize the expected total cost of the restoration activities. The design variables are the idle time periods and the restorations of the individual bridges in the network. The proposed approach was applied to an existing bridge network in Santa Barbara, California, under a strong earthquake. This bridge network was represented by 11 nodes, 28 road segments, and 38 individual bridges. Using the computer program HAZUS (FEMA 2009), a fragility analysis at network-level was performed, and the three objectives were formulated. A Pareto optimal solution indicates the restoration schedule and restoration cost allocated to each individual bridge, the network functionality over time along with the expected total restoration cost over time, and restoration cost allocation to all individual bridges, as shown in Figures 8.13(a) and 8.13(b).

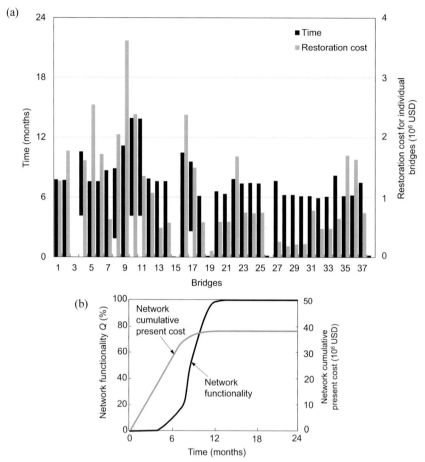

Figure 8.13 Optimum bridge network restoration: (a) restoration schedule and restoration cost of individual bridges; (b) time-dependent functionality and cumulative cost for restoration activities (adapted from Bocchini and Frangopol 2012b).

A general approach for life-cycle management of deteriorating civil infrastructure systems considering resilience to lifetime extreme events was presented by Yang and Frangopol (2019a). Their study introduces a novel concept of lifetime resilience loss of a deteriorating structure subjected to multiple extreme events during its life. The lifetime resilience of a deteriorating structure is characterized by its cumulative losses to lifetime hazards. The lifetime hazards and life-cycle performance are modeled as renewal-reward processes. The lifetime resilience loss RS_{loss}^{life} is estimated as (Yang and Frangopol 2019a)

$$RS_{loss}^{life} = \sum_{i=1}^{N_{ex}(t_{life})} RS_{loss,i} \qquad (8.18)$$

where $N_{ex}(t_{life})$ is the number of extreme events during lifetime t_{life}, and $RS_{loss,i}$ is the resilience loss due to the ith extreme event. Figure 8.14 illustrates the lifetime resilience losses of a bridge under progressive and sudden deterioration. The performance of a deteriorating bridge after each extreme event can be improved and recovered by applying minimal repair and replacement actions. As indicated in Yang and Frangopol (2019a), minimal repairs are defined as the immediate corrective actions conducted after the occurrence of a hazard. This type of repairs is only able to reverse the direct detrimental effect of the hazard and do not address any issues posed by progressive deterioration. Replacement actions are taken whenever the performance is lower than a prescribed threshold. Finally, major repairs are scheduled in advance to counteract adverse effects from progressive deterioration by enhancing the structural performance to its initial level. The resilience loss associated with a minimal repair is related to both performance losses induced by continuous deterioration and extreme events. The resilience loss associated with a replacement is only affected by the performance threshold for replacement and the recovery time, as shown in Figure 8.14. The formulations of the expected lifetime intervention costs, risks, and resilience losses are based on the renewal theory. The multi-objective optimization based on minimizing the expected lifetime intervention costs, expected lifetime failure risk, and expected lifetime resilience loss is used for optimum intervention planning. This planning indicates the major repair schedule and/or the performance threshold for replacement. It was concluded that (a) lifetime intervention costs, lifetime failure risks, and lifetime resilience losses can be significantly underestimated by ignoring continuous deterioration; (b) a limited number of proactive major repairs and timely corrective actions after extreme events can reduce the lifetime intervention costs, lifetime failure risks, and lifetime resilience losses; and (c) the consideration of progressive deterioration and minimal repairs is necessary in the life-cycle management of deteriorating structures under extreme events.

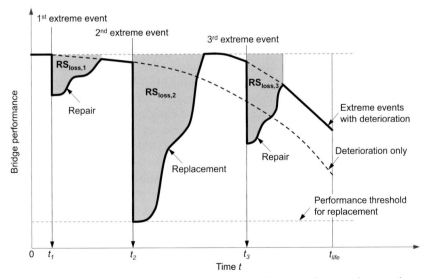

Figure 8.14 Resilience losses with maintenance interventions under continuous deterioration and extreme events (adapted from Yang and Frangopol 2019a).

8.4   SUSTAINABILITY OF BRIDGES AND BRIDGE NETWORKS

According to Brundtland (1987), sustainable development is meeting the needs of the present without compromising the ability of future generations to meet their own needs. The American Society of Civil Engineers (ASCE 2017a) defines sustainability "as a set of economic, environmental and social conditions in which all of society has the capacity and opportunity to maintain and improve its quality of life indefinitely without degrading the quantity, quality or the availability of economic, environmental and social resources." The concept of sustainability has been adopted in various ways depending on the application fields (Bocchini et al. 2014). Sustainability has been recognized as an appropriate performance indicator for bridge management under hazards over the last decade.

8.4.1   General Concept of Sustainability for Bridge Management

Sustainability typically considers both direct and indirect consequences due to hazards. These consequences are assessed in terms of social, environmental, and economic aspects (Lee and Lin 2008; Dong et al. 2013, 2014a; Yadollahi et al. 2015). The social component of sustainability is estimated considering travel time delayed caused by bridge failure. The environmental component should address an increase in CO_2 emissions induced by the changed traffic flow and running on detours. The economic considerations include the costs associated with constructing damaged or failed bridges and fatalities (Dong et al. 2014b; Liu et al. 2018). Bridges and bridge networks can be considered sustainable if the cost and energy spent CO_2 emissions associated with maintenance interventions are less than their target levels. The sustainable bridge management under hazards is related to maximizing the service life, minimizing life-cycle cost, and minimizing social and environmental impacts under the required performance criteria (Bocchini et al. 2014; Zoubir and McAllister 2016). A unified risk-based approach addressing resilience and sustainability of civil infrastructure simultaneously was presented in Bocchini et al. (2014). To bridge the gap between sustainability and resilience, Yang and Frangopol (2018a) introduced the concept of lifetime resilience and proposed an integrated framework for probabilistic life-cycle optimization of civil infrastructure considering both sustainability and resilience.

Life-cycle performance profiles associated with unsustainable and sustainable bridge design are illustrated in Figure 8.15(a). The bridge performance profiles with more and less sustainable maintenance interventions after an extreme event are compared in Figure 8.15(b), where resilience values of two profiles are the same since the recovery processes including recovery time, rapidity and profile are identical. As shown in Figure 8.15, the bridge design and maintenance considering sustainability can result in a larger initial performance, larger damage initiation time, less performance deterioration rate, and larger service life than the bridge design without sustainability.

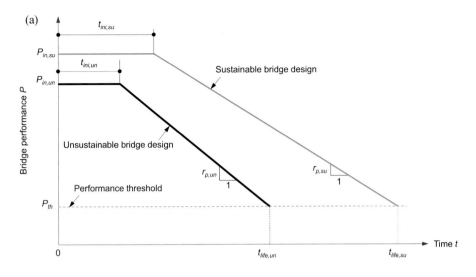

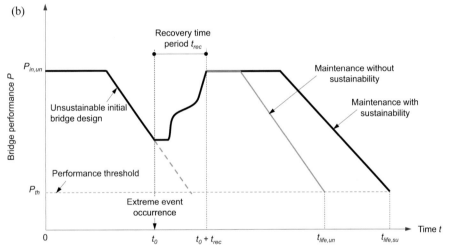

Figure 8.15 Bilinear life-cycle bridge performance profiles considering sustainability: (a) unsustainable and sustainable bridge designs; (b) non-sustainable and sustainable maintenance interventions.

8.4.2 Sustainability Assessment

Sustainability can be assessed as the sum of product of the conditional consequences and the associated probabilities of occurrence of events that cause these consequences. The consequences related to social, environmental, and economic impact due to the failure of single and multiple bridges in a network are converted into monetary values. For an individual bridge, the monetary loss $C_{SU}^{IN}(t)$ in terms of sustainability at time t can be formulated as (Dong et al. 2014b)

$$C_{SU}^{IN}(t) = \sum_{i=1}^{N_{ex}} \sum_{j=1}^{N_{ls}} \sum_{k=1}^{N_{ds}} \left[C_{con|ds,k} \cdot \left\{ P_{ds,k|ls,j}(t) \cdot P_{ls,j|ex,i}(t) \cdot P_{ex,i}(t) \right\} \right] \quad (8.19)$$

where N_{ex}, N_{ls}, and N_{ds} are the numbers of extreme events (e.g, earthquakes, hurricanes), structural limit states (e.g., yielding, buckling), damage states (e.g., no damage, slight damage, moderate damage, and extensive) to be considered, respectively, $C_{con|ds,k}$ is the consequence conditioned by the kth damage state, $P_{ds,k|ls,j}(t)$ is the conditional occurrence probability of the kth damage state given the jth structural limit state at time t, $P_{ls,j|ex,i}(t)$ is the occurrence probability of the jth structural limit state conditioned by the ith extreme event at time t, and $P_{ex,i}(t)$ is the occurrence probability of the ith extreme event at time t. The subsequent section provides information on the estimation of monetary loss in terms of sustainability of bridge networks.

Social Loss

Under seismic hazards, the social metric of sustainability includes the extra travel time for the users due to bridge damages and failures in a network. The expected extra travel time $E[t_{etr}(t)]$ in a bridge network at time t can be estimated as (Dong et al. 2014a; Dong and Frangopol 2019)

$$E\left[t_{etr}(t)\right] = \sum_{i=1}^{N_l} \sum_{j=1}^{N_{ds}} P_{lds,ij|GI}(t) \cdot t_{etr,ij} \quad (8.20)$$

where N_l is the number of links in the bridge network, N_{ds} is the number of damage states, $P_{bds,ij|GI}(t)$ is the probability that the ith link is in the jth damage state due to an earthquake with a specific ground motion intensity at time t, and $t_{etr,ij}$ is the extra travel time of the ith link for the jth damage state. Considering that the bridge damages and failures in a network lead to an increase in travel time and distance, the extra travel time $t_{etr,ij}$ in Eq. (8.20) can be computed as (Decò et al. 2013; Dong et al. 2014b; Dong and Frangopol 2019)

$$t_{etr,ij} = t_{d,ij} \left[ADR_{ij} \left(\frac{d_i}{S_{dmg,i}} - \frac{d_i}{S_{0,i}} \right) + ADD_{ij} \cdot \frac{d_{dt,i}}{S_{dt}} \right] \quad (8.21)$$

where $t_{d,ij}$ is the downtime associated with the jth damage state of the ith link (days), d_i is the length of link i(km), $S_{dmg,i}$ and $S_{0,i}$ are the traffic speeds (km/h) when the ith link is damaged and intact, respectively, ADR_{ij} and ADD_{ij} are the average daily traffic remaining and average daily traffic detoured at the ith link in damage state j, respectively, $d_{dt,i}$ is the length of the detour for the ith link (km), and $S_{dt,i}$ is the detour speed for the ith link (km/hr).

The extra travel time $t_{etr,ij}$ results in monetary loss for the users and goods. Therefore, the expected total social monetary loss of the network $E[C_{SC}(t)]$ can be expressed as (Decò et al. 2013; Dong et al. 2014a)

$$E\left[C_{SC}\left(t\right)\right] = \sum_{i=1}^{N_l}\sum_{j=1}^{N_{ds}} P_{lds,ij|GI}\left(t\right)\cdot C_{etr,ij} \tag{8.22}$$

The monetary loss associated with the jth damage state of the ith link $C_{etr,ij}$ is computed as (Decò et al. 2013; Dong et al. 2014a, 2014b)

$$C_{etr,ij} = \left[c_w \cdot O_{car}\cdot\left(1 - r_{AADT}\right) + \left(c_{tc}\cdot O_{trk} + c_{gds}\right)\cdot r_{AADT}\right]\cdot t_{etr,ij} \tag{8.23}$$

where c_w is the wage per hour (USD/h), c_{tc} is the total compensation per hour (USD/h), c_{gds} is the time value of goods transported (USD/h), O_{car} and O_{trk} are the vehicle occupancies for cars and trucks, respectively, and r_{AADT} is the ratio of average daily truck traffic to average daily traffic. The extra travel time $t_{etr,ij}$ can be estimated as indicated in Eq. (8.21).

Environmental Loss

The environmental loss of a bridge network under seismic hazards is computed considering that the traffic detour on a link after an earthquake can lead to an increase in travel time, produce additional carbon dioxide emissions, and require additional energy. Moreover, repair actions can result in energy waste and carbon dioxide emissions. The expected bridge network environmental metric related to the traffic detour after an earthquake $E[W_{EN,tr}(t)]$ at time t can be computed as (Dong et al. 2013, 2014a)

$$E\left[W_{EN,tr}\left(t\right)\right] = \sum_{i=1}^{N_l}\sum_{j=1}^{N_{ds}}\left\{\begin{array}{l} P_{lds,ij|GI}\left(t\right)\cdot ADT_{ij}\left(t\right)\cdot d_i\cdot t_{etr,ij} \\ \cdot\left[EN_{car}\cdot\left(1 - r_{AADT}\right) + EN_{trk}\cdot r_{AADT}\right]\end{array}\right\} \tag{8.24}$$

where EN_{car} and EN_{trk} are the environmental metrics per unit distance (e.g., carbon dioxide ton/km) for cars and trucks, respectively, and ADT_{ij} is the average daily traffic at the ith link in damage state j. The repair actions are determined according to the damage states induced by earthquakes. Depending on the repair actions, energy waste and carbon dioxide emissions can be estimated. Therefore, the expected bridge network environmental metric due to energy waste $E[W_{EN,ew}(t)]$ at time t can be expressed as (Dong et al. 2014a)

$$E\left[W_{EN,ew}\left(t\right)\right] = \sum_{i=1}^{N_{bdg}}\sum_{j=1}^{N_{ds}}\left\{\begin{array}{l} P_{bds,ij|GI}\left(t\right)\cdot r_{c,ij} \\ \cdot\left[EN_{steel}\cdot V_{steel,i} + EN_{conc}\cdot V_{conc,i}\right]\end{array}\right\} \tag{8.25}$$

in which N_{bdg} is the number of bridges in a network, $P_{bds,ij|GI}(t)$ is the probability that the ith bridge is in the jth damage state after an earthquake with a certain ground motion intensity at time t, $r_{c,ij}$ is the repair cost ratio for the ith bridge at damage state j, EN_{steel} and EN_{conc} are the environmental metrics per unit volume (e.g., carbon dioxide ton/m^3) for steel and concrete, respectively, and $V_{steel,i}$ and $V_{conc,i}$ are the volumes of steel and concrete of the ith bridge, respectively. Since $E[W_{EN,tr}(t)]$ of Eq. (8.24) and $E[W_{EN,ew}(t)]$ of Eq. (8.25) are based on a weight unit, the expected total bridge network environmental monetary loss $E[C_{EN}(t)]$ is expressed as (Dong et al. 2014b)

$$E\left[C_{EN}\left(t\right)\right] = \left\{E\left[W_{EN,tr}\left(t\right)\right] + E\left[W_{EN,ew}\left(t\right)\right]\right\} \cdot c_{EN} \qquad (8.26)$$

where c_{EN} is the unit cost of the environmental metric (e.g., USD/ton).

Economic Loss

The economic loss of sustainability includes the costs associated with bridge repair loss and life loss (Dong et al. 2013, 2014b; Liu et al. 2018). The expected repair and construction cost for damaged or failed bridges $E[C_{EC,rp}(t)]$ at time t is expressed as (Stein et al. 1999; Dong et al. 2013; Dong and Frangopol 2019)

$$E\left[C_{EC,rp}\left(t\right)\right] = \sum_{i=1}^{N_{bdg}} \sum_{j=1}^{N_{ds}} P_{bds,ij|GI}\left(t\right) \cdot r_{c,ij} \cdot c_{rp} \cdot A_i \qquad (8.27)$$

where c_{rp} = repair (or construction) cost per unit area (e.g., USD/m^2); A_i = repaired (or constructed) area of bridge i. Furthermore, assuming that fatalities are caused by bridge failure following an earthquake, the expected life loss induced by an earthquake $E[C_{EC,fi}(t)]$ at time t can be estimated as (Decò et al. 2013; Dong et al. 2014a, 2014b; Dong and Frangopol 2019)

$$E\left[C_{EC,fi}\left(t\right)\right] = \sum_{i=1}^{N_{bdg}} \sum_{j=1}^{N_{ds}} P_{bds,ij|GI}\left(t\right) \cdot FT_{ij} \cdot C_{imp} \qquad (8.28)$$

where FT_{ij} is the number of fatalities of the ith bridge being in the jth damage state, and C_{imp} is the implied cost of a fatality. The expected total economic monetary loss $E[C_{EC}(t)]$ is

$$E\left[C_{EC}\left(t\right)\right] = E\left[C_{EC,rp}\left(t\right)\right] + E\left[C_{EC,fi}\left(t\right)\right] \qquad (8.29)$$

Finally, the total monetary loss in terms of sustainability $E[C_{SU}^{NET}(t)]$ can be obtained as

$$E\left[C_{SU}^{NET}\left(t\right)\right] = E\left[C_{SC}\left(t\right)\right] + E\left[C_{EN}\left(t\right)\right] + E\left[C_{EC}\left(t\right)\right] \qquad (8.30)$$

The values of $E[C_{SC}(t)]$, $E[C_{EN}(t)]$ and $E[C_{EC}(t)]$ are computed using Eqs. (8.22), (8.26) and (8.29), respectively.

Time-Dependent Sustainability

Figure 8.16 illustrates the time-dependent profiles associated with the expected total monetary loss for sustainability metric $E[C_{SU}^{NET}(t)]$, expected social loss $E[C_{SC}(t)]$, expected environmental loss $E[C_{EN}(t)]$, and expected economic loss $E[C_{EC}(t)]$ of a bridge network under seismic events assuming no maintenance. The network, consisting of 15 bridges and 5 nodes, is located in the San Francisco Bay Region (Dong et al. 2014a). These expected losses increase over time due to the deterioration of the seismic performance of bridges in the network. In general, both the expected social monetary loss $E[C_{SC}(t)]$ and environmental monetary loss $E[C_{EN}(t)]$ are much larger than the expected economic monetary loss $E[C_{EC}(t)]$.

In order to assess time-variant sustainability under seismic events as shown in Figure 8.16, the computational procedure consists of four components (Dong et al. 2014a): (a) identification of the seismic activity of the region; (b) bridge network configuration; (c) consequence analysis; and (d) sustainability analysis. The possible seismic events should be identified for the regions where the bridge network is located. The bridge network configuration includes bridge inventory and network information such as dimensions, location, type and year built of individual bridges, and average daily traffic involved in the network, among others. The consequence analysis quantifies social, environmental, and economic monetary losses. As indicated in Eqs. (8.20) to (8.29), these losses require the fragility analysis, which provides the conditional probability that the bridge response caused by various levels of ground shaking exceeds the structural capacity defined by a damage state, and outcome estimations (e.g., downtime, energy wastes, carbon dioxide emissions, repair cost, and fatalities) resulted from each damage state of all the individual bridges in a network. Finally, the total monetary loss in terms of sustainability can be obtained. Since the bridge performance deteriorates over time (see Section 2.3), the fragility analysis is time-dependent. Figure 8.17(a) compares the fragility curves at time t_0, t_1, and t_2. It can be seen that the probability of exceeding a certain damage state for a given ground motion intensity increases with time. As a result, the time-dependent consequences and total monetary loss in terms of sustainability can be predicted as shown in Figure 8.16.

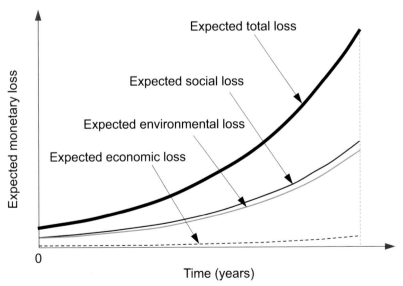

Figure 8.16 Time-variant contributions of various types of losses to the expected total loss of a bridge network in terms of sustainability without maintenance intervention over time (adapted from Dong et al. 2014a).

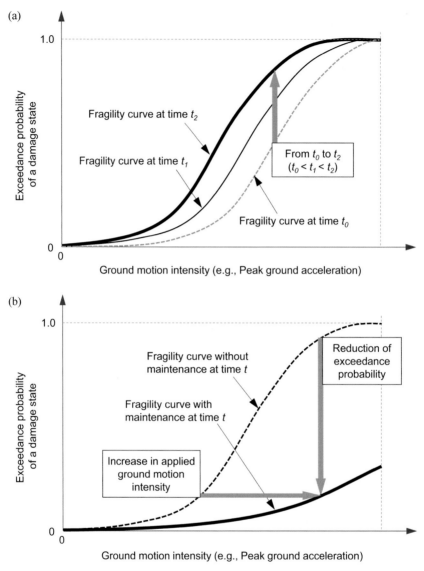

Figure 8.17 Comparison of fragility curves of deteriorating bridges: (a) fragility curves at different times; (b) fragility curves with and without maintenance.

Sustainability with Maintenance

The maintenance actions can affect the fragility curve of a deteriorating bridge. Figure 8.17(b) compares the fragility curves for deteriorating bridges with and without maintenance interventions. As shown in this figure, the improvement of the bridge performance by applying maintenance actions leads to (a) a reduction of the probability of exceeding a certain damage state for a given ground motion intensity, and (b) an increase of the ground motion intensity associated with a probability of exceeding a certain damage state. When a maintenance action is applied on a bridge in the ith damage state at time t, the median of ground motion intensity $m[G_{i,m}(t)]$ can be expressed as (Dong et al. 2014b, 2015)

$$m\left[G_{i,m}\left(t\right)\right] = m\left[G_{i,0}\left(t\right)\right]\cdot\left(1 + r_{ma,i}\cdot R_{lev}\right) \tag{8.31}$$

where $m[G_{i,0}(t)]$ is the median of ground motion intensity of the bridge without maintenance at time t, $r_{ma,i}$ is the enhancement ratio of ground motion for a bridge in the ith damage state after maintenance, and R_{lev} is the retrofit level, which ranges from 0 to 1.0. The retrofit action leading to fully strengthen the seismic performance is associated with $R_{lev} = 1.0$.

8.5 APPLICATIONS OF SUSTAINABILITY FOR OPTIMUM BRIDGE NETWORK MANAGEMENT

Bridge network management under seismic events can be optimized to minimize the monetary loss based on sustainability and determine the optimum times and types of maintenance. Recently, bridge management based on sustainability has been treated as an effective tool for seismic mitigation of bridges and bridge networks (Dong and Frangopol 2019). This section presents several developed approaches and their case studies on sustainability assessment and sustainability-based bridge network management.

8.5.1 Applications of Sustainability Assessment for Individual Bridges and Bridge Networks

The optimum bridge management under seismic events can be formulated based on the sustainability assessment. Dong et al. (2013) developed an approach to assess the time-variant sustainability of individual bridges under seismic events. The seismic fragility over time was computed based on nonlinear finite element analysis of a bridge, where the effects of corrosion-induced deterioration and flood-induced scour on seismic fragility were taken into account. The proposed approach was applied to a typical single bent and two-span RC bridge, designed based on Caltrans' bridge design specification and Seismic Design Criteria (CSDC 2004). The time-dependent fragility curves associated with only corrosion and with both corrosion and flood-induced scour were compared. The time-dependent expected economic loss with and without considering flood-induced scour was quantified. Finally, it was concluded that (a) the ground motion intensity significantly affects the sustainability, (b) the corrosion and flood-induced scour result in a severe reduction in structural capacity of bridges, (c) the effects of corrosion and flood-induced scour on the metrics of sustainability increase over time, and (d) the variation in downtime of a bridge caused by an earthquake significantly affects the economic metric of sustainability.

Based on the approach presented in Dong et al. (2013), a framework for the time-variant sustainability of highway bridge networks under seismic hazard was

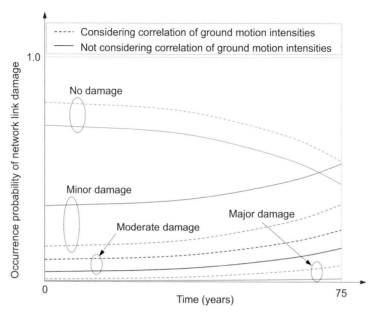

Figure 8.18   Time-variant occurrence probabilities of network link damage index with and without considering the correlations among the ground motion intensities at different sites under a seismic scenario (adapted from Dong et al. 2014a).

introduced in Dong et al. (2014a). In this framework, the time-variant fragility curve of each bridge is used to compute the network link damage index and associated occurrence probability. Based on the network link damage index, the link flow capacity and the flow speed are determined. Finally, the sustainability of highway bridge network considering the expected costs of repair, running of detouring vehicles, time loss, environmental loss, and life loss is estimated. A bridge network located in the San Francisco Bay Region was adopted for a case study. This bridge network consists of 5 nodes and 15 individual bridges. The individual bridges are categorized into three types (i.e., (Type A) simply supported concrete bridge with multiple column bents, (Type B) multiple span continuous concrete bridge, and (Type C) single-span concrete bridge). The fragility curves of three types of bridges were obtained over the time horizon of 75 years. The correlations among the ground motion intensities at different sites under a given seismic scenario were considered to compute the network link damage index. The comparison between the occurrence probabilities of network link damage index with and without considering the correlations among the ground motion intensities is presented in Figure 8.18. From this figure, it is found that there is a difference between the occurrence probabilities with and without considering the correlations among the ground motion intensities, and the occurrence probability for no link damage decreases, but the occurrence probabilities for minor, moderate and major link damages increase over time. It was concluded that (a) sustainability is time-variant due to both bridge performance deterioration and discount rate of money, (b) the sustainability analysis should consider the ground motion correlation of spatially distributed systems, and (c) the parameters used in cost analysis and the relation between the link and bridge performance affect the accuracy of sustainability metrics, and therefore, should be estimated carefully.

8.5.2 Optimal Sustainability-Based Management of Bridges and Bridge Networks

A probabilistic optimum pre-earthquake retrofit planning for bridge networks based on sustainability was presented in Dong et al. (2014b). The sustainability was quantified in terms of the expected economic losses by converting both the social metric (i.e., downtime and fatalities) and the environmental metric (i.e., energy waste and carbon dioxide emissions) into a monetary value. This approach formulates the bi-objective optimization problem by simultaneously minimizing both the total retrofit cost for the bridge network and the expected economic loss for an investigated time horizon. By solving the bi-objective optimization problem, the Pareto front, which corresponds to the optimum sequence and timing of bridge retrofitting, was obtained. The proposed approach was applied to an existing highway bridge network consisting of 10 concrete bridges located in Orange County, California. The time effects on the fragility curves of bridges with and without retrofit and links under probabilistic seismic scenarios are accounted for. It was concluded that (a) including sustainability in probabilistic management of bridge networks is essential, (b) prioritization of bridges to be retrofitted within a network can be performed based on a probabilistic multi-objective optimization approach, (c) the economic metric of sustainability increases in time rapidly as a result of bridge deterioration, (d) the retrofit planning depends on the allocated budget (e.g., a higher budget can result in an optimum retrofit planning associated with a smaller maximum expected annual risk), (e) the time interval between retrofit actions significantly affects the optimum planning (e.g., a longer time interval between retrofit actions can produce a higher risk level), and (f) the performance of highway bridge networks under seismic hazard can be improved by using the proposed approach.

Dong et al. (2015) developed a framework for the seismic retrofit optimization of highway bridge networks based on cost-benefit analysis and multi-attribute utility associated with sustainability. The developed bi-objective optimization framework consists in simultaneously maximizing the utility associated with the seismic retrofit cost of the bridge network and the utility associated with the benefit of seismic retrofit actions within a prescribed time horizon. The utility associated with seismic retrofit cost U_C is formulated considering the total cost of retrofit of a highway bridge network as (Ang and Tang 1984)

$$U_C = \frac{1}{1-\exp(-\gamma)}\left[1-\exp\left(\frac{-\gamma\left(C_{R,max}-C_R\right)}{C_{R,max}}\right)\right] \qquad (8.32)$$

where $C_{R,max}$ is the maximum retrofit cost, C_R is the total expected retrofit cost, and γ is the parameter to represent the risk attitude of the decision maker. The positive and negative values of γ are associated with risk averse and risk accepting attitudes, respectively. The multi-attribute utility associated with a bridge network sustainability is expressed as (Jiménez et al. 2003)

$$U_{S,0} = w_{SC} \cdot U_{SC} \cdot E(C_{SC,0}) + w_{EN} \cdot U_{EN} \cdot E(C_{EN,0}) + w_{EC} \cdot U_{EC} \cdot E(C_{EC,0}) \qquad (8.33a)$$

$$U_{S,R} = w_{SC} \cdot U_{SC} \cdot E(C_{SC,R}) + w_{EN} \cdot U_{EN} \cdot E(C_{EN,R}) + w_{EC} \cdot U_{EC} \cdot E(C_{EC,R}) \qquad (8.33b)$$

where $U_{S,0}$ and $U_{S,R}$ indicates the multi-attribute utilities without and with retrofit, respectively. w_{SC}, w_{EN} and w_{EC} are the weight factors, and U_{SC}, U_{EN} and U_{EC} are the utility functions. The subscripts SC, EN and EC in the weight factors and utility functions are associated with the social, environmental, and economic metrics, respectively. The benefit utility due to the retrofit actions during a period of T years can be expressed as

$$U_B(T) = U_{S,R}(T) - U_{S,0}(T) \qquad (8.34)$$

Dong et al. (2015) defines the cost-benefit indicator CB associated with the retrofit actions as

$$CB(T) = U_B(T) - (1 - U_C) \qquad (8.35)$$

The cost-benefit indicator CB represents the effectiveness of a bridge network retrofit plan, and ranges from –1 and 1. CB larger than zero shows that the retrofit plan is beneficial.

In the developed framework by Dong et al. (2015), the bi-objective optimization is performed to find the type of retrofit action performed on each bridge of the network at the beginning of the time interval considered, such that both the utility associated with the retrofit cost of the bridge network and the utility associated with the retrofit benefit corresponding to the time interval considered are maximized. This framework is applied to an existing highway bridge network located in Alameda, California, which consists of fifteen individual bridges. Three retrofit options with the retrofit levels equal to 0.25, 0.5, and 1.0 are considered, where the retrofit level equal to 1 results in a fully restoring seismic performance of a bridge. Through the bi-objective optimization described previously, the two Pareto fronts associated with time intervals of 20 and 30 years are shown in Figure 8.19(a). A single Pareto optimal solution in this figure indicates the type of retrofit action for all the individual bridges for a given time interval (i.e., 20 or 30 years), and a risk averse attitude (i.e., $\gamma = 2$) of the decision maker. By comparing the Pareto fronts associated with time intervals of 20 and 30 years, it can be found that for the same utility associated with total retrofit cost, the time interval of 20 years produces a lower utility associated with benefit. Figure 8.19(b) compares the two Pareto fronts based on risk averse (i.e., $\gamma = 2$) and accepting (i.e., $\gamma = -2$) attitudes of a decision maker when a 30 year time interval is adopted. For the same utility associated with the total retrofit cost, a risk accepting attitude of a decision maker leads to a smaller benefit utility than that associated with a risk averse attitude. Therefore, the time interval under investigation and the risk attitude of a decision maker significantly affect the Pareto front.

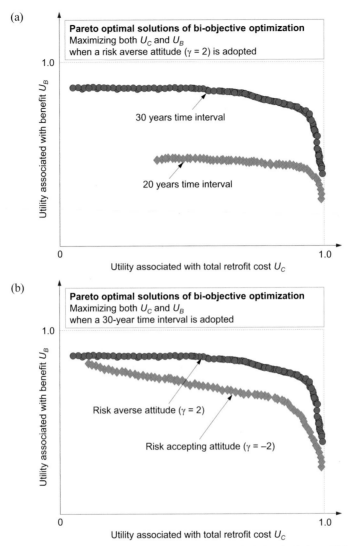

Figure 8.19 Comparison of Pareto fronts based on maximizing both the utilities associated with the retrofit cost and benefit: (a) 20 and 30 years time intervals; (b) risk averse and accepting attitudes of a decision maker (adapted from Dong et al. 2015).

A sustainability-informed bridge ranking approach for a network of bridges vulnerable to scour was investigated by Liu et al. (2018). This approach uses multi-attribute utility theory and transportation network analysis. The proposed ranking method using the expected sustainability utility integrates both bridge condition and bridge importance, and considers the economic, social, and environmental consequences of bridge failure. The proposed approach was illustrated on an existing highway bridge network located in Camden County, New Jersey. The results reported indicate that the proposed approach is able to identify the bridges with severe substructure condition and high-network-level importance. Therefore, a rational allocation of the available funds and maintenance interventions on a highway network of bridges subjected to potential scour hazard can be achieved.

8.6   CONCLUDING REMARKS

This chapter deals with resilience and sustainability for management of individual bridges and bridge networks under extreme events. The general concept of resilience for bridge management is described. The effects of bridge performance deterioration and hazard-induced damage, and risk mitigation on resilience are addressed. Since resilience depends on the recovery process under uncertainty, an appropriate functionality recovery model should be determined. For this reason, several representative recovery models are provided. Applications of resilience assessment for individual bridges and bridge networks are summarized. Recent investigations and associated applications for optimal resilience-based management of bridges and bridge networks are provided. Furthermore, in order to consider the economic, social, and environmental consequences of bridge failures in a holistic manner, the concept of sustainability is presented, and approaches for sustainability assessment and optimum bridge network management based on sustainability are described. Applications related to sustainability assessment of bridges and bridge networks and sustainability-based bridge network management are provided. The management of bridges considering climate change is addressed in Chapter 9.

Chapter 9

Bridge Management Considering Climate Change

A_u	= annual amplification coefficient for scale parameter
A_α	= annual amplification coefficient for shape parameter
A_λ	= annual amplification coefficient for occurrence rate of hurricanes
C_0	= chloride concentration at surface
$C_{0,i}$	= present value of the cost of the retrofit for the ith adaptation strategy if it were applied at times t_0
C_A	= total expected adaptation cost
C_B	= benefit by applying adaptation actions
C_{crt}	= crtical chloride concentration
$C_{fld,i}$	= expected loss due to the ith flood occurring at time t_i
$C_{h,i}$	= expected loss due to the ith hurricane occurring at time t_i
$C_{G,i}$	= gain of the ith adaptation strategy
$C_{t,i}$	= present value of the cost of the retrofit applied at the time t^* specified in the ith adaptation strategy
$C(x,\ t)$	= chloride concentration at depth x at time t
$C_{T,fld}\ (t_{int})$	= total flood loss during the time interval $(0,\ t_{int})$
$C_{T,h}\ (t_{int})$	= total loss due to hurricanes during the time interval $(0,\ t_{int})$
D_c	= chloride diffusion coefficient
$F_V(v,\ t)$	= time-varying cumulative distribution function of the wind speed during hurricane
g_{hur}	= performance function for the bridge deck unseating failure mode due to hurricane
g_{LT}	= performance function associated with lateral failure mode
g_{VT}	= performance function associated with vertical failure mode
$N_f(t_{int})$	= number of flood occurrence during the time interval $(0,\ t_{int})$
$N_h(t_{int})$	= number of hurricane occurrence during the time interval $(0,\ t_{int})$
$P_{dmg}(t)$	= probability of severe corrosion damage occurrence at time t
$P_{f,fld}$	= probability of failure induced by flood

NOTATIONS (continued)	
$P_{in}(t)$	= probability of corrosion initiation at time t
$P(E_{fld,a})$	= annual probability of flood occurrence
$P(F|E_{fld,a})$	= conditional failure probability given the flood occurrence
$P(N=N_h)$	= probability that N_h-time hurricanes will occur during the time interval $(0, t_{int})$
$Q_{T,c}$	= flood intensity
r_{bc}	= benefit-cost ratio
$r_{cor}(t)$	= corrosion rate at time t
r_{dmg}	= damage ratio
r_{mag}	= parameters representing the change in the magnitude of flood
r_{frq}	= parameters representing the change in the occurrence frequency of flood
$r_{gl,i}$	= gain-loss ratio for the ith adaptation strategy
$R_{0,i}^{an}$	= average annual risks associated with the retrofit option for the ith adaptation strategy applied at time t_0
$R_{t,i}^{an}$	= average annual risks associated with the retrofit option for the ith adaptation strategy applied at time t^*
$R_{k,0}$	= total life-cycle cost without adaptation
$R_{k,a}$	= total life-cycle cost with adaptation
$R_{K,fld}(t)$	= annual risk of bridge scour under flood
t_{sc}	= time of severe corrosion damage occurrence
T_c	= flood occurrence time interval
$x_c(t)$	= carbonation depth $x_c(t)$ at time t

ABSTRACT

Chapter 9 describes the effects of climate change on bridge performance under uncertainty, adaptation measures for bridge management, cost-benefit analysis, and optimum adapations for individual bridges and bridge network management. Corrosion, flood-induced scour and hurricane threats with or without considering climate change, and their effects on bridge performance indicators and service life are described. Several investigations associated with the prediction of bridge network performance under climate change are provided. Furthermore, representative adaptation measures and their effects on bridge performance are described. The benefit-cost analysis to optimize the time and types of adaptation measures for bridge management is presented.

9.1   INTRODUCTION

Climate change has been one of the critical concerns for bridge management. Climate change may lead to more intense and more frequent floods, hurricanes, and other climate-related hazards, as well as changes in CO_2 atmospheric concentration, temperature, and humidity. As a result, the bridge performance can be considerably reduced in a short-term period (Bastidas-Arteaga et al. 2010, 2013; IPCC 2014; Wang et al. 2012; Stewart and Deng 2015). The increased vulnerability of bridges under uncertain climate change poses a significant challenge to bridge managers (Mondoro et al. 2018a). Therefore, significant efforts to investigate bridge management considering climate change have been made recently (Akiyama et al. 2020).

In this chapter, the effect of climate change on bridge performance, adaptation measures for bridge management, and optimum adaptation strategy for individual bridges and bridge network management are presented. Climate change can lead to changes in environmental factors affecting corrosion initiation and propagation, and magnitude and occurrence frequency of hazards. Accordingly, the results of several investigations associated with the prediction of bridge and bridge network performance under climate change are provided. Furthermore, several representative adaptation measures and their effects on bridge performance are described. The benefit-cost analysis used for decision making to determine the time and types of adaptation measures for bridge management is presented. For more informed decision making, the optimization of adaptation strategy considering climate change is described, where the objectives based on the benefit-cost and gain-loss analysis can be applied.

9.2 BRIDGE PERFORMANCE UNDER CLIMATE CHANGE

Change in the climate is related to increases in CO_2 atmospheric concentration, occurrence rate of natural hazards (e.g., earthquake, hurricane, flood, and tsunamis), and their magnitude (Stewart et al. 2011; Stewart and Deng 2015; Dong and Frangopol 2016b). The probability density functions (PDFs) of the magnitude of a natural hazard with and without climate change are compared in Figure 9.1(a). This figure indicates that climate change can increase the mean and standard deviation of the magnitude of the hazard. The cumulative distribution functions (CDFs) associated with the PDFs in Figure 9.1(a) are illustrated in Figure 9.1(b). These CDFs correspond to the probability that the magnitude of the natural hazard does not exceed a predefined value. This figure shows that (a) climate change can produce a higher probability of exceeding a predefined value of magnitude $M_{n,0}$ (i.e., $P_{exd,c} > P_{exd,0}$, where $P_{exd,c}$ and $P_{exd,0}$ are the exceedance probabilities associated with climate change and no climate change, respectively), and (b) climate change can result in a larger magnitude of the natural hazard for a given probability of occurrence $P_{M,n}$ than that associated with no climate change (i.e., $M_{n,c} > M_{n,0}$, where $M_{n,c}$ and $M_{n,0}$ are the magnitudes of the natural hazard for the given probability $P_{M,n}$ associated with climate change and no climate change, respectively). Rational considerations of the changes in magnitude and occurrence probability of hazard due to climate change and their effects on bridge performance under uncertainty are required for effective bridge performance management under climate change (Mondoro et al. 2018a; Akiyama et al. 2020).

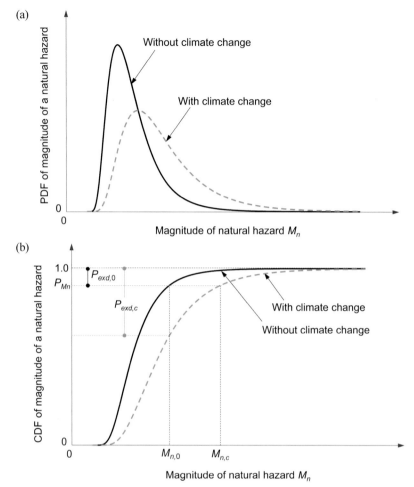

Figure 9.1   Magnitude of a natural hazard with and without climate change: (a) probability density functions (PDFs); (b) cumulative distribution functions (CDFs).

9.2.1   Impact of Climate Change on Performance of RC Structures under Corrosion

Corrosion in RC structures is affected by various environmental factors including temperature, humidity, and the concentration of CO_2 in the atmosphere (Bastidas-Arteaga and Stewart 2015; Stewart et al. 2011, 2012). Stewart et al. (2011, 2012) and Wang et al. (2012) assessed the probabilities of corrosion initiation and corrosion damage for existing RC civil infrastructure in Australia. Their studies estimated the changes in carbonation depth, chloride penetration, corrosion initiation time, and probability of corrosion damage of RC structures under climate change between 2000 and 2100. Considering the CO_2 concentration and temperature over time, the carbonation depth $x_c(t)$ at time t can be expressed as (Stewart et al. 2011)

$$x_c(t) = \sqrt{\frac{2 \cdot f_T(t) \cdot D_{CO_2}(t)}{k_a} k_{urb} \int_{2000}^{t} M_{CO_2}(t)dt \cdot \left(\frac{1}{t-1999}\right)^{k_{ag}}} \quad \text{for } t \geq 2000 \quad (9.1)$$

where $f_T(t)$ is the function representig the temperature effect on diffusion coefficient; $D_{CO_2}(t)$ is the CO_2 diffusion coefficient in concrete at time t; k_a is the factor characterizing the cement; k_{urb} is the factor for the increased CO_2 levels in urban environments; $M_{CO_2}(t)$ is the mass concentration of ambient CO_2 at time t; and k_{ag} is the age factor. Figure 9.2(a) shows the increase in CO_2 concentration and temperature under climate change over time. Accordingly, the mean carbonation depth $E[x_c(t)]$ also increases over time under climate change.

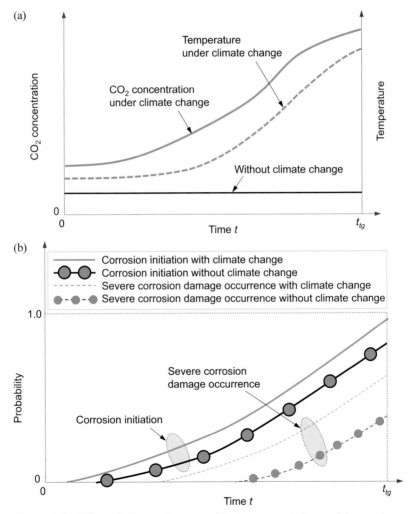

Figure 9.2 Effect of climate change on (a) CO_2 concentration and temperature; (b) probabilities of corrosion initiation and severe corrosion damage occurrence (adapted from Stewart et al. 2011).

Based on Fick's law, the time-dependent chloride concentration $C(x, t)$ at depth x can be estimated as (Stewart et al. 2011)

$$C(x,t) = C_0 \left[1 - erf \left(\frac{x}{2 \left(\dfrac{k_{en} \cdot k_{ts} \cdot k_{cr} \cdot f_T(t)}{\cdot D_c \cdot t_0^n (t-1999)^{1-n}} \right)^{0.5}} \right) \right] \quad \text{for } t \geq 2000 \quad (9.2)$$

where C_0 is the chloride concentration at surface, k_{en} is the environmental factor, k_{ts} is the test method factor, k_{cr} is the curing factor, D_c is the chloride diffusion coefficient, t_0 is the curing time (i.e., 0.0767 years), and n is the aging factor. The relation between corrosion rate $r_{cor}(t)$ and temperature $Tmp(t)$ at time t is expressed as (DuraCrete 2000)

$$r_{cor}(t) = r_{cor-20} \left[1 + K \cdot (Tmp(t) - 20) \right] \quad (9.3)$$

where r_{cor-20} is the corrosion rate at 20°C, and K is the corrosion rate coefficient, where $K = 0.025$ for $Tmp(t) < 20°C$, and $K = 0.073$ for $Tmp(t) \geq 20°C$. Using the corrosion rate, the severe corrosion damage occurrence time can be estimated.

Based on Eqs. (9.1) and (9.2), the probability of corrosion initiation $P_{in}(t)$ at time t is computed as (Stewart et al. 2011; Wang et al. 2012)

$$P_{in}(t) = P(x_0 - x_c(t) < 0) \quad (9.4a)$$

$$P_{in}(t) = P(C_{crt} - C(x_0, t) < 0) \quad (9.4b)$$

where $x_0 = $ depth associated with reinforcement bar; and $C_{crt} = $ critical chloride concentration for corrosion initiation. The probability of severe corrosion damage occurrence $P_{dmg}(t)$ at time t is (Stewart et al. 2011; Wang et al. 2012)

$$P_{dmg}(t) = P(t \geq t_{sc}) \quad (9.5)$$

where t_{sc} is the severe corrosion damage occurrence time, which is based on the three stages (i.e., corrosion initiation, corrosion crack initiation, and crack propagation to severe cracking). Figure 9.2(b) shows the evolution of the time-dependent probability of corrosion initiation and the probability of severe corrosion damage occurrence with and without climate change. Stewart et al. (2011, 2012) and Wang et al. (2012) concluded that (a) there will be a significant carbonation-induced and chloride-induced corrosion in RC structures for specific locations in Australia; (b) an increase in CO_2 concentration and changes in both temperature and humidity due to climate change will accelerate corrosion damage in RC structures and reduce their service life; and (c) the increase in the probability of chloride-induced damage occurrence is less than the increase in the probability of carbonation-induced damage occurrence.

9.2.2 Impact of Climate Change on Bridge Performance under Flood-Induced Scour

Flood-induced scour can significantly reduce the performance of bridges by exposing their pier foundations. For the flood-induced scour analysis, extensive information related to flood magnitude and frequency is required. For given magnitude and occurrence frequency of flood, probabilistic bridge performance can be estimated considering both vertical and lateral failure modes of pier foundations (Dong and Frangopol 2016b). The performance functions g_{VT} and g_{LT} associated with vertical and lateral failure modes can be expressed, respectively, as (Dong and Frangopol 2016b)

$$g_{VT} = \rho_{R,VT} \cdot r_{dmg} \cdot R_{VT} - \rho_{L,VT} \cdot L_{VT} \tag{9.6a}$$

$$g_{LT} = \rho_{R,LT} \cdot r_{dmg} \cdot R_{LT} - \rho_{L,LT} \cdot L_{LT} \tag{9.6b}$$

where $\rho_{R,VT}$ and $\rho_{L,VT}$ are the unbiased values of resistance and load effect for the vertical failure mode, respectively, r_{dmg} is the damage ratio ($r_{dmg} = 0$ for full damaged state, and $r_{dmg} = 1$ for no damage state), and R_{VT} and L_{VT} indicate the vertical resistance and load effect, respectively. The subscript LT is associated with lateral failure mode. Considering both the vertical and lateral failure modes, the probability of failure induced by flood $P_{f,fld}$ is computed as

$$P_{f,fld} = P\left[\left(g_{VT} \leq 0\right) \cup \left(g_{LT} \leq 0\right)\right] \tag{9.7}$$

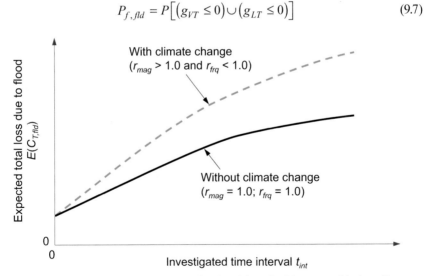

Figure 9.3 Expected total loss due to floods with and without considering climate change (adapted from Dong and Frangopol 2016b).

Climate change can be addressed in the formulation of the flood-induced probability of failure. The flood intensity and occurrence time interval considering climate change are used to compute the vertical and lateral load effects of the performance functions [i.e., L_{VT} and L_{LT} in Eq. (9.6)], and the associated probability of failure $P_{f,fld}$ of Eq. (9.7). The flood intensity $Q_{T,c}$ and occurrence time interval T_c under climate change are expressed as (Dong and Frangopol 2016b)

$$Q_{T,c} = r_{mag} \times Q_T \qquad (9.8a)$$

$$T_c = r_{frq} \times T \qquad (9.8b)$$

where r_{mag} and r_{frq} are the parameters representing the changes in the magnitude and occurrence frequency of flood, respectively. When no climate change is considered, r_{mag} and r_{frq} are equal to one. The negative effect of climate change on bridge performance can be represented by r_{mag} larger than one (i.e., larger flood intensity), and r_{frq} smaller than one (i.e., more frequent flood occurrence). The expected total flood loss with and without climate change is compared in Figure 9.3. The total flood loss $C_{T,fld}(t_{int})$ is computed as (Yeo and Cornell 2005)

$$C_{T,fld}\left(t_{int}\right) = \sum_{i=1}^{N_f\left(t_{int}\right)} C_{fld,i} \cdot e^{-r_{dis} \cdot t_i} \qquad (9.9)$$

where $N_f(t_{int})$ is the number of flood occurrences during the time interval from 0 to t_{int}, $C_{fld,i}$ is the expected loss due to bridge failure caused by the ith flood occurring at time t_i, and r_{dis} is the discount rate of money. The expected loss $C_{fld,i}$ consists of repair loss, operation loss and loss caused by travel time delay due to bridge failure (Dong and Frangopol 2016b). Figure 9.3 shows that climate change (i.e., $r_{mag} > 1.0$ and $r_{frq} < 1.0$) leads to a larger expected total flood loss than that associated with no climate change.

The flood vulnerability of a two-span continuous box girder bridge under three flood scenarios associated with 100, 200 and 500 years was investigated by Dong and Frangopol (2016b) by computing the scour depth under each flood scenario according to Briaud et al. (1999). The expected life-cycle flood loss under the three different flood scenarios considered and the comparison of expected life-cycle loss considering climate change under different hazard intensities and frequencies have been reported in Dong and Frangopol (2016b). Their study revealed that (a) the time-dependent performance deterioration and hazards significantly affect the resilience and total life-cycle loss of bridges, (b) the total life-cycle loss is sensitive to changes in the indirect consequences due to bridge failure, time from the occurrence of the last hazard, discount rate of money, and remaining service life; and (c) since different hazards could dominate the life-cycle loss, specific risk mitigation strategies associated with various hazards have to be determined.

Climate change can be considered for estimating the flood-induced risk (Yang and Frangopol 2019a). Annual risk of bridge under flood-induced scour $R_{K,fld}(t)$ can be computed as (Ellingwood and Kinali 2009; Mondoro et al. 2017)

$$R_{K,fld}(t) = P(E_{fld,a}) \cdot P(F|E_{fld,a}) \cdot C_{fail}(t) \qquad (9.10)$$

where $P(E_{fld,a})$ is the annual probability of flood occurrence, $P(F|E_{fld,a})$ is the conditional failure probability given the flood occurrence, and $C_{fail}(t)$ is the present value of consequences due to bridge failure at time t. The probabilities $P(E_{fld,a})$ and $P(F|E_{fld,a})$ are affected by flood frequency and magnitude under climate change, respectively (Yang and Frangopol 2019b). Figure 9.4 shows the PDFs of annual maximum flow discharge due to flood under climate change. These PDFs can be obtained using the flow discharge data. The area under the PDF above the discharge of scour-critical flood Q_{fld} corresponds to $P(E_{fld,a})$. The probability $P(F|E_{fld,a})$ is the ratio of the area under the PDF of annual maximum flow discharge above the minimum discharge resulting in failure Q_{fail} to $P(E_{fld,a})$. This figure shows that an increase in the flow discharge results in increasing both probabilities $P(F|E_{fld,a})$ and $P(E_{fld,a})$, and finally annual risk $R_{K,fld}(t)$ of Eq. (9.10) increases.

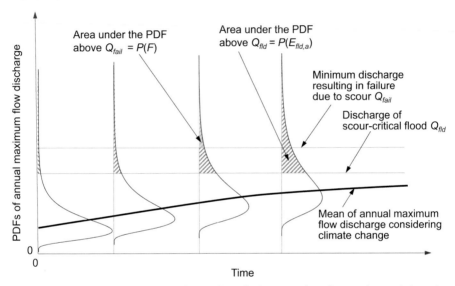

Figure 9.4 PDFs of annual maximum flow discharge under climate change (adapted from Yang and Frangopol 2019b).

A probabilistic approach to evaluate the climate change impact on the risk of bridge scour, including hydrologic modeling based on geospatial data collected from several public databases, was developed by Yang and Frangopol (2019b). The proposed approach, applied to bridges in the Lehigh River watershed, integrates global climate models with future climate change scenarios to estimate the long-term regional risk of bridge scour. Their findings include (a) quantification of uncertainties arising from different global climate models is essential for the analysis and planning of effective climate adaptation strategies; (b) information on bridge foundation is essential for accurate bridge scour risk analysis; (c) downscaled climate data based on global climate model can be used to estimate the changes in the temperature of the specific region; (d) depending on the case study, the climate change scenario associated with a drastic temperature increase may result in decreasing in flood frequency and scour risk; and (e) climate change is more likely to affect flood frequency than flood scour hazard intensity.

9.2.3   Impact of Climate Change on Bridge Performance under Hurricanes

The low-clearance single-span bridges are susceptible to deck unseating due to uplift forces generated by severe storms and hurricanes (Mondoro et al. 2017). The total vertical uplift force is computed as the sum of the maximum quasi-static force and the associated vertical slamming force (AASHTO 2007b). Therefore, the performance function for the bridge deck unseating failure mode g_{hur} is expressed as (Mondoro et al. 2017)

$$g_{hur} = (R_w + R_{yld}) - (L_{v,qs} + L_{v,sl}) \qquad (9.11)$$

where R_w and R_{yld} are the weight per unit length of the deck and the yield strength of the tie-downs, respectively. $L_{v,qs}$ and $L_{v,sl}$ indicate the maximum quasi-static vertical force and vertical slamming force, respectively. The computations of $L_{v,qs}$ and $L_{v,sl}$ require the wind speeds and surge heights induced by hurricanes.

The time-varying CDF of the wind speed $F_V(v, t)$ during a hurricane can be represented by the Weibull distribution as (Mondoro et al. 2017; Dong and Frangopol 2017)

$$F_V\left(v, \ t\right) = 1 - \exp\left[-\left(\frac{v}{u(t)}\right)^{\alpha(t)}\right] \qquad (9.12)$$

where v = wind speed; $u(t)$ = scale parameter at time t; and $\alpha(t)$ = shape parameter at time t. These two parameters considering climate change can be expressed as

$$u(t) = u_0 + A_u \cdot t^{r_u} \qquad (9.13a)$$

$$\alpha(t) = \alpha_0 + A_\alpha \cdot t^{r_\alpha} \qquad (9.13b)$$

where u_0 and α_0 are the scale and shape parameters without considering climate change, respectively, A_u and A_α are the annual amplification coefficient for u and α considering climate change effects, and r_u and r_α are the changing rates associated with u and α, respectively. If hurricane occurrence follows a Poisson distribution, the probability that N_h hurricanes will occur during the time interval $(0, t_{int})$ is expressed as (Ang and Tang 2007)

$$P\left(N = N_h\right) = \frac{\left(\lambda_0 \cdot t_{int}\right)^{N_h}}{N_h!} \cdot \exp\left(-\lambda_0 \cdot t_{int}\right) \qquad (9.14)$$

where λ_0 is the mean occurrence rate of hurricanes. The mean occurrence rate of hurricanes at time t associated with climate change can be computed as

$$\lambda\left(t\right) = \lambda_0 + A_\lambda \cdot t^{r_\lambda} \qquad (9.15)$$

where A_λ is the annual amplification coefficient, and r_λ is the changing rate. By replacing λ_0 with $\lambda(t)$, climate change can be implemented to compute the probability of N_h hurricane occurrences provided by Eq. (9.14). Furthermore, the total loss due to hurricanes $C_{T,h}(t_{int})$ is computed as (Yeo and Cornell 2005; Dong and Frangopol 2016b, 2017)

$$C_{T,h}\left(t_{int}\right) = \sum_{i=1}^{N_h\left(t_{int}\right)} C_{h,i} \cdot e^{-r_{dis} \cdot t_i} \qquad (9.16)$$

where $N_h(t_{int})$ is the number of hurricane occurrences during the time interval $(0, t_{int})$, and $C_{h,i}$ is the expected loss due to the ith hurricane occurring at time t_i. Figure 9.5 shows the effects of the annual amplification coefficients A_u and A_λ on the expected total loss. The values of A_u and A_λ larger than 1.0 indicate the increases in the wind speed and hurricane occurrence rate, respectively. These increases produce an increase in the expected total loss as shown in Figure 9.5.

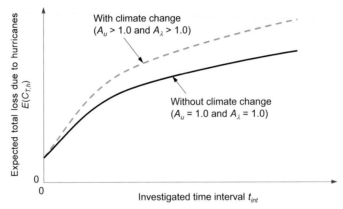

Figure 9.5   Expected total loss due to hurricanes with and without considering climate change (adapted from Dong and Frangopol 2017).

9.2.4   Bridge Network Performance under Climate Change

The bridge network performance estimation is based on the performance indicators (e.g., reliability, probability of failure, life-cycle cost) of individual bridges as described in Chapter 7. By adopting the probabilistic approaches for performance prediction of individual bridges under climate change, the bridge network performance under climate change can be predicted. However, it is challenging to predict the bridge network performance under climate change, because the associated computation requires (a) integrating information on climate change, (b) addressing the effect of climate change on the performance deteriorations of multiple individual bridges in the network, and frequency and intensity of possible hazards, and (c) dealing with uncertainties associated with bridge network performance under climate change effectively and efficiently. Therefore, extensive investigations to estimate the bridge network performance under climate change are needed (Mondoro et al. 2018a; Akiyama et al. 2020).

Recently, a generalized approach to assess the societal risk of transportation networks considering climate change and future population growth was developed by Yang and Frangopol (2019c). This approach addresses the effects of (a) climate change on structural deterioration and flood frequencies and (b) population growth on traffic demand of transportation networks under uncertainty. The proposed methodology can consider different components of the transportation networks, such as pavements and bridges, and different threats to these networks induced by climate change, such as flooding, hurricanes and corrosion. The robust traffic flow patterns of transportation networks under uncertainty are obtained using a robust traffic assignment model based on the expected residual minimization approach. These flow patterns are used to compute the societal risk. The generalized approach was illustrated on a real-world transportation network located in Camden County, New Jersey, under uncertainties due to climate change and population growth. Both flood and traffic hazards are considered in assessing the vulnerability of the transportation network. Figure 9.6 shows the annual risk of transportation network under climate change and population growth. Climate change and population growth lead to bridge performance deterioration with an increase in annual occurrence probability of extreme floods and annual increase in traffic demand. In addition to the computational efficiency of the proposed methodology, their study revealed that (a) climate change and population growth can significantly increase the societal risk; (b) extreme floods affected by climate change can dramatically alter traffic flow patterns of the network; (c) an increase in traffic demand due to population growth results in an increase in the network traffic volume; and (d) risk associated with time losses accounts for a greater portion of the total societal risk than that associated with running losses.

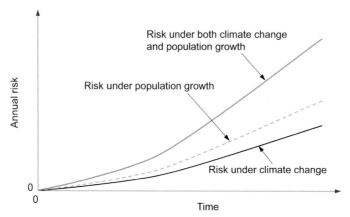

Figure 9.6   Annual risk of transportation network under climate change and population growth (adapted from Yang and Frangopol 2019c).

9.3   ADAPTATION FOR BRIDGE MANAGEMENT UNDER CLIMATE CHANGE

The effect of climate change on bridge performance can be mitigated through appropriate adaptation strategies for bridge management (IPCC 2014; Stewart and Deng 2015). The adaptation strategies include enhancement of bridge design standards, utilization of new materials, and retrofitting and strengthening of bridges (Bastidas-Arteaga and Stewart 2015). The retrofitting and strengthening of bridges improve the resistance against deck unseating, substructure failure, coastal erosion, wind damage, and scour, among others (Mondoro et al. 2018a).

Extreme uplift and transverse hydraulic forces induced by hurricanes can damage the bridge deck (Ataei and Padgett 2013). In order to reduce these forces on a bridge deck and increase its resistance (a) large potential storm surge inundation and wave height can be adopted (Rosenzweig et al. 2011); (b) holes in a bridge superstructure can be inserted to reduce the buoyancy force acting on the superstructure during floods and hurricanes (Sawyer 2008); and (c) connections between the bridge superstructure and substructure (e.g., tie-down, restrainers, and anchorage bars) can be applied to increase the resistance against upward and transverse movement (Okeil and Cai 2008; Ataei and Padgett 2013).

The shear and flexural capacities of a bridge substructure may be insufficient to resist hydraulic loads during extreme hydraulic events. When the continued exposure to chlorides corrodes the RC bridge substructures, the shear and flexural capacities of the bridge decrease dramatically. Depending on the degree of damage, maintenance (e.g., surface coatings, pile wraps, pile jackets with reinforcement) and complete replacement can be applied to restore the shear and flexural capacities (FHWA 2010; Mondoro et al. 2018a). Scour at the bridge substructure is caused by extreme precipitation, which reduces the capacity and stability of the bridge substructure. Scour countermeasures can prevent damage in the bridge substructures (Mondoro et al. 2018a). These countermeasures include approach-channel control, downstream-channel control, armoring of the bridge opening, and drainage control (NCHRP 2007; Agrawal et al. 2007).

Extreme winds associated with severe coastal hurricanes can pose a significant threat to the safety of long-span bridges. During extreme winds, excessive vibration resulting in high stress ranges in critical locations (e.g., cable anchorages) can cause fatigue crack initiation and propagation (Li et al. 2002; Fujino and Siringoringo 2013; Zhang et al. 2014). Wind-induced vibration can be reduced by adopting vibration control systems. The fatigue cracks propagation can be prevented by applying the appropriate inspections and maintenances as indicated in Sections 3.2 and 4.2. A detailed review on adaptation methods for bridge management under climate change can be found in Mondoro et al. (2018a).

9.4 BENEFIT-COST ANALYSIS FOR BRIDGE MANAGEMENT

In order to optimize the bridge adaptation strategies under climate change, benefit-cost analysis (or cost-benefit analysis) is essential (Dong and Frangopol 2019). The benefit for bridge management under climate change is defined as the reduction in total life-cycle cost or risk by applying adaptation actions, which is expressed as (Dong and Frangopol 2017; Mondoro and Frangopol 2018)

$$C_B = R_{k,0} - R_{k,a} \qquad (9.17)$$

where C_B is the benefit by applying adaptation actions, and $R_{k,0}$ and $R_{k,a}$ are the risks associated with the bridge without and with adaptation, respectively. The benefit-cost ratio r_{bc} is defined as (Mondoro and Frangopol 2018)

$$r_{bc} = \frac{C_B}{C_A} \qquad (9.18)$$

where C_A is the total expected adaptation cost. An adaptation strategy with a benefit-cost ratio r_{bc} less than one indicates that the adaptation is not cost-effective. The ratio r_{bc} larger than one denotes that it is beneficial to perform the adaptation. Since the risks $R_{k,0}$ and $R_{k,a}$ are affected by the deterioration of bridge performance, the benefit-cost ratio r_{bc} varies with the investigated time period. Therefore, an adaptation measure may be cost-ineffective considering a short period of the investigated time, but cost-effective if associated with a long period of the investigated time. Considering the time-dependent damage propagation under uncertainty, the probability that an adaptation measure is cost-effective can be estimated as $P(r_{bc} > 1.0)$ (Bastidas-Arteaga and Stewart 2015).

A risk-based benefit-cost analysis is investigated to determine the retrofit strategies for bridges under extreme hydrologic events (e.g., flooding, hurricanes, tsunamis) by Mondoro and Frangopol (2018). Their study evaluated the benefit-cost ratios based on the risk associated with deck failure, foundation failure, and failure of an entire riverine bridge vulnerable to flooding. The failure of the bridge is defined as the failure of a series system with three possible failure modes including deck failure, pier failure or foundation failure. The risk associated with the bridge failure is computed considering the consequences of each possible failure mode. Five types of adaptation measures were considered: (a) riprap to increase the resistance of the foundation against scour; (b) steel restrainers to reduce the deck displacement in the transverse and uplift directions; (c) shear keys to restrain the transverse displacement; (d) riprap and restrainers, and (e) riprap and shear keys. The benefit-cost ratios for three flood hazard exposure cases represented by discharging following the log-Pearson type III distribution were estimated. The pier failure probability and the risk associated with this event were found to be negligible. Figure 9.7 shows the probabilities of failure, risks and benefit-cost ratios of deck, foundation, and entire bridge for the five types of adaptation measures considered under the less severe flood hazard exposure case. From Figure 9.7, it can be seen that (a) all retrofit options considered reduce the probability of failure of the bridge, but only some result in the decrease in risk [see Figures 9.7(a) and 9.7(b)]; (b) adaptation measures to prevent deck dislodgement (i.e., steel restrainers and shear keys) reduce the probability of failure of the bridge deck and entire bridge, but lead to an increase in the probability of failure of the foundation [see Figure 9.7(a)]; (c) if only deck failure mode is considered, a high benefit-cost ratio can be obtained by applying steel restrainers (or shear keys), but if the failure mode of entire bridge is considered, the application of steel restrainers (or shear keys) is not cost-effective [see Figure 9.7(c)]. The benefit-cost ratios associated with the entire bridge considering the five types of adaptation

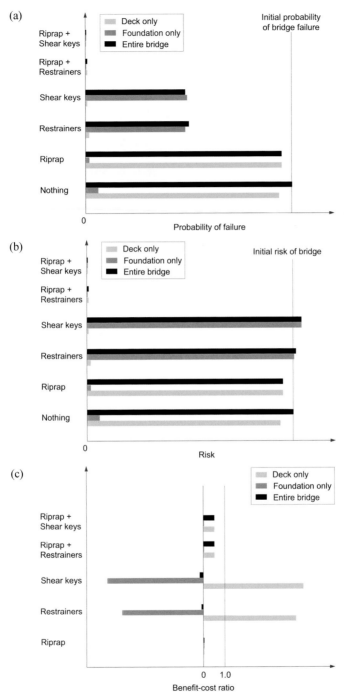

Figure 9.7 Effects of adaptation measures under flood hazards on (a) probabilities of failure; (b) risks; (c) benefit-cost ratio associated with deck, foundation, and entire bridge (adapted from Mondoro and Frangopol 2018).

measures under the three flood hazard exposure cases considered (i.e., Cases A, B and C) are shown in Figure 9.8. This figure shows that under moderate- and large-scale flood exposure conditions (i.e., Cases B and C) the largest benefit-cost ratio for the entire bridge is related to the adaptation measures associated with riprap and restrainer, and riprap and shear keys.

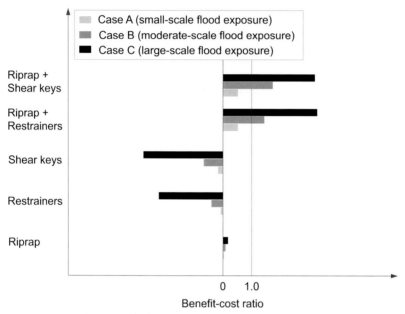

Figure 9.8   Benefit-cost ratio for the bridge considering adaptation measures under three flood exposure cases (adapted from Mondoro and Frangopol 2018).

9.5   OPTIMUM ADAPTATION OF BRIDGE MANAGEMENT

Several approaches to optimize the adaptation types and times for bridge management under climate change have been investigated recently. Mondoro et al. (2018b) developed a bi-objective optimization approach considering climate change hazard by using robust optimization for identifying optimal climate change adaptation strategies in order to address deep uncertainties. The benefit-cost and gain-loss ratios are the proposed objectives to be optimized simultaneously. The gain of the ith adaptation strategy $C_{G,i}$, defined as the present value of interest earned by delaying adaptation, is expressed as

$$C_{G,i} = C_{0,i} - C_{t,i}$$
(9.19)

where $C_{0,i}$ is the present value of the cost for the retrofit option involved in the ith adaptation strategy if it was applied at time t_0, and $C_{t,i}$ is the present value of the cost of the retrofit option applied at the time t^* specified in the ith adaptation strategy. The difference between the average annual risks associated with the retrofit option involved in the ith adaptation strategy, if it was applied at time t_0 (denoted as $R_{0,i}^{an}$), and the retrofit option applied at the time t^*, specified in the ith adaptation strategy (denoted as $R_{t,i}^{an}$), is referred to as the loss associated with the delay $t^* - t_0$, which is expressed as

$$C_{L,i} = R_{0,i}^{an} - R_{t,i}^{an} \qquad (9.20)$$

The gain-loss ratio for the ith adaptation strategy $r_{gl,i}$ is (Mondoro et al. 2018b)

$$r_{gl,i} = \frac{C_{G,i}}{C_{L,i}} \qquad (9.21)$$

This ratio is a metric used to systematically assess the effect of delaying the adaptation actions on the economic gain and risk. The benefit-cost ratio and gain-loss ratio represent the economic efficiency and flexibility of the adaptation strategy, respectively.

Using the benefit-cost ratio and gain-loss ratio, Mondoro et al. (2018b) developed three types of bi-objective optimizations based on pessimistic and optimistic approaches. The pessimistic approach considers the worst climate scenario, which minimizes the benefit-cost ratio r_{bc} and gain-loss ratio r_{gl}. Therefore, the associated bi-objective optimization formulation consists in finding the types and times of adaptations which maximize both the minimum r_{bc} and minimum r_{gl} for a set of climate scenarios. The optimistic bi-objective optimization considers the best climate scenario resulting in the maximum benefit-cost and gain-loss ratios. The objectives are maximizing both the maximum r_{bc} and maximum r_{gl} for a set of climate scenarios. The third bi-objective optimization proposed relies on considering the probability of occurrence of each climate change scenario assumed, and consist in the maximization of both the benefit-cost and gain-loss ratios. The design variables and given conditions are the same in all bi-objective optimizations considered. Illustrative examples associated with a typical bridge over two rivers with different potential climate change trends considering flooding are presented. Flooding is consistently expected to become more intense in one river and less intense in the other.

The conclusions of Mondoro et al. (2018b) include (a) Pareto optimal solutions of the pessimistic bi-objective optimization indicate that there will be no gain by delaying adaptation under the worst climate change scenario, (b) need to prevent significant losses outweighs the desire for the high gain-loss ratio for the regions associated with an overall intensification of the hazard, and (c) desire for the high gain-loss ratio can lead to optimal adaptation strategies allowing to delay the adaptation for the regions associated with an overall decrease in the intensity of the hazard.

A risk-based framework to determine the optimal adaptation schedules for bridge networks vulnerable to long-term flood-induced scour risk under climate change was developed by Liu et al. (2020). This framework requires (a) evaluation of the consequences of bridge failure at network-level, (b) estimation of bridge scour under possible climate change scenarios, and (c) investigation of effects of budget availability and risk perception on the decision making. The adaptation optimization is formulated with the objective of minimizing the total life-cycle cost including adaptation cost and agency and user costs associated with network damage scenarios. The design variables are the time of retrofit application for all bridges, and the constraints include the annual budget for adaptation and the number of retrofit actions during a specified time horizon. The highway bridge network vulnerable to abutment scour illustrated in their study consists of 10 bridges and 66 intersections located in the Lehigh River watershed in Pennsylvania. Riprap, a widely available and cost-effective scour countermeasure to armor bridge abutments, is applied as the retrofitting measure. The annual risks of bridge network with and without adaptation actions are compared in Figure 9.9. It shows that the application of short-term retrofit actions (e.g., riprap) can reduce the long-term scour risk.

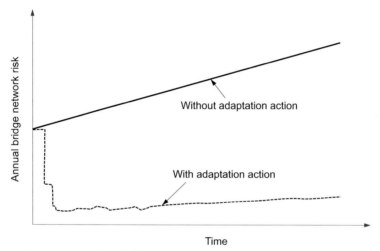

Figure 9.9   Annual risk of bridge network under climate change with and without adaptation actions (adapted from Liu et al. 2020).

The approach proposed by Liu et al. (2020) revealed that (a) optimal adaptation schedules for all bridges in a network under long-term flood-induced scour risk can be made by considering regional climate prediction and local hydrologic condition, (b) adequacy of the budget and necessity of supplementary appropriation can be determined, (c) decision makers can effectively provide the optimum climate adaptation strategy for bridge network management, and (d) network-level risks and the associated benefit-cost ratios can be assessed considering the decision maker's perception of user cost.

9.6 CONCLUDING REMARKS

This chapter deals with the effects of climate change on bridge performance under uncertainty, adaptation measures for bridge management, cost-benefit analysis, and optimum adaptations for individual bridges and bridge network management.

Corrosion, flood-induced scour and hurricane threats with or without climate change, and their effects on bridge performance indicators (e.g., probabilities of corrosion initiation and severe corrosion damage occurrence, probability of failure, total loss, and annual risk) and service life are described. Several recent investigations on bridge network performance considering climate change are presented, based on the probabilistic performance prediction of individual bridges. Also, representative adaptation measures to reduce the impact of climate change on bridge performance are provided. The concepts of benefit-cost analysis and gain-loss analysis, and their applications are presented. It is shown that optimal risk-based adaptation strategies of bridge networks under climate change can take into consideration the cost perception of decision makers and provide the adaptation types and times for each bridge in the network.

Chapter **10**

Conclusions

ABSTRACT

This chapter provides (a) the summary of this book, (b) representative conclusions of each chapter, and (c) future efforts in the research fields of life-cycle bridge safety, maintenance, and management in a life-cycle context.

10.1 SUMMARY

This book is designed to cover the fundamental and recently developed concepts and approaches related to life-cycle bridge safety, maintenance and management under uncertainty. The main topics include bridge safety and service life prediction, bridge inspection and maintenance, life-cycle bridge and bridge network management, resilience and sustainability of bridges and bridge networks under hazards, and bridge management considering climate change. Chapter 1 addresses the general concepts and methods of bridge life-cycle performance and cost analysis. Chapter 2 presents the probabilistic time-dependent structural performance and service life prediction of deteriorating bridges. Representative techniques of inspection and SHM and their use for bridge management are addressed in Chapter 3. Chapter 4 deals with the effects of bridge maintenance on performance, service life, and cost. Based on these effects, life-cycle bridge management can be optimized. In Chapter 5, the probabilistic concepts and approaches for optimum life-cycle bridge management planning are described, including the objectives for optimum life-cycle bridge management, multi-objective optimization, and multi-attribute decision making. Chapter 6 presents applications of optimum life-cycle bridge management planning with emphasis on probabilistic performance and service life prediction, optimum inspection and monitoring planning, and optimum maintenance planning at a single bridge-level (i.e., project-level). Chapter 7 provides the general concepts of the bridge network performance, optimum life-cycle bridge management at a network-level, and their applications. Chapter 8 presents the resilience and sustainability for the management of bridges and bridge networks under extreme events. The impact of climate change on the management of bridges at both project-level and network-level is the topic of Chapter 9.

10.2 CONCLUSIONS

The following representative conclusions are drawn:

- Bridge maintenance and management involve uncertainties. These uncertainties increase as the bridge performance and cost assessment are predicted further into the future. Therefore, probabilistic concepts and methods are necessary to identify these uncertainties and to reduce the epistemic uncertainties as much as possible. Updating based on inspection and monitoring information is one of the processes used to reduce some uncertainties involved in bridge maintenance and management.

- Due to the scarcity of financial resources allocated for maintaining and/ or improving the performance of bridges, optimization of the resources available is crucial. The significance of using multi-objective optimization in bridge maintenance management under uncertainty has to be emphasized when making decisions regarding requirements for continued service of deteriorating bridges and bridge networks.

- The optimum bridge management planning considers various probabilistic objectives based on the life-cycle bridge performance assessment and prediction, and cost evaluation processes. Probabilistic simulations and complex conditions to address the uncertainties associated with parameters and models involved in these processes are necessary. For this reason, a high level of computational efficiency for life-cycle bridge performance and cost assessment and prediction is essential for optimum bridge management planning.

- The optimum bridge management planning can be based on various objectives considering the life-cycle performance and cost of individual bridges and bridge networks. The objectives for optimum bridge management planning can be categorized into: (a) performance-based objectives, (b) cost-based objectives, (c) damage detection-based objectives, (d) service life-based objectives, and (e) risk-based objectives. Increasing the number of objectives considered for bridge maintenance planning can lead to more rational and well-balanced solutions through multi-objective optimization. However, an increase in the number of objectives results in more difficulties associated with the computation to obtain the Pareto front and decision-making to select well-balanced solutions.

- The life-cycle optimal bridge management at a network-level needs to be adopted to improve the efficiency and effectiveness of life-cycle management of all bridges in the network under limited financial resources.

Approaches considering life-cycle optimizations at the project-level and at the network-level into one mathematical problem are also providing decisions makers with a range of practically implementable solutions.

- The bridge network maintenance planning provides the optimum times and types of inspection and maintenance for all bridges in a network. This planning can be optimized with various bridge network performance-based objectives. Their formulations need to estimate the reliability, connectivity, travel time and distance, maintenance cost, user cost, failure cost, risk, sustainability, and resilience of the bridge network. These estimations need the life-cycle performance and cost analysis of each bridge in the network.

- The concepts of resilience and sustainability are helpful to manage the life-cycle risk of the bridge network. The risk management of bridge networks based on resilience and sustainability can be optimized to maximize the bridge network resilience, to minimize both the expected total cost and time for the restoration activities, to minimize the time required to reach a target functionality level, and to minimize the monetary loss associated with sustainability.

- Climate change poses a substantial threat to bridges by increasing their probability of unsatisfactory performance and failure due to the increase in the intensity and frequency of floods and hurricanes, sea level rise, higher scour rates, and accelerated material degradation, among others. Bridge management has to consider the impact of climate change by (a) integrating the prediction of future climate and converting it to structural demand, (b) quantifying the uncertainties, (c) investigating the effect of adaptation measures on the bridge performance under climate change, and (d) optimizing the adaptation strategies considering bridge performance, risk and cost-benefit.

10.3   FUTURE DIRECTIONS

There are still further developments to be addressed as follows:

- Appropriate probabilistic modeling of bridge performance deterioration requires a realistic and generalized damage propagation modeling considering bridge types, materials used in the bridge components, environmental conditions, and loading effects. Effective and practical methods for capturing bridge performance deterioration under no maintenance and under different maintenance interventions and loading scenarios in a life-cycle context will continue to present an important challenge.

- In general, life-cycle cost and risk analyses for bridges and bridge networks are highly affected by both direct cost (e.g., maintenance, repair, and replacement costs) and indirect cost (e.g., economic, social, and environmental losses). A balanced minimization of the direct and indirect costs provides a useful means to generate solutions for optimum long-term bridge network performance and expenditures. A reliable life-cycle cost analysis has to consider the uncertainties in predicting lifetime bridge performance at project and network levels. Further efforts in this direction are required for reliable life-cycle cost and risk analyses of bridges and bridge networks.

- The optimum bridge maintenance planning is the process to determine the optimum times and types of inspection, monitoring, and maintenance considering multiple objectives. Although an increase in the number of objectives of the optimization process leads to more rational and well-balanced solutions, there will be difficulties concerning computation cost and decision making. For this reason, efficient algorithms to improve the ability to search the Pareto front in order to detect the most promising solutions considering the risk attitude of the decision makers should be developed.

- The uncertainties associated with the optimum safety and maintenance planning for individual bridges are unavoidable, since the life-cycle bridge performance and cost analyses are highly uncertain. Furthermore, since the bridge network management under progressive deterioration due to aggressive environments and sudden deterioration due to extreme events requires extensive information on the performance of individual bridges, modeling of bridge network, progressive deterioration and hazard occurrence prediction, the uncertainties associated with this bridge network management will become larger. For this reason, further investigations to quantify the uncertainties involved in bridge network management in a life-cycle context are necessary.

- When the appropriate probabilistic concepts and methods are adopted to integrate the information related to bridge network management, epistemic uncertainty can be reduced, but aleatory uncertainty remains (Der Kiureghian and Diltlevsen 2009). Some of the uncertainties introduce interdependencies among the bridges of the network and among the hazards. Incorporation of these interdependencies in the optimization process of bridge network management under multiple hazards is a field in its infancy.

- During the past two decades, advanced probabilistic techniques and approaches for life-cycle bridge and bridge network management have been developed. However, there are still difficulties in the practical application of these techniques and approaches. These difficulties are related to the limited data available on the performance of individual bridges, effects of maintenance interventions on bridge performance, progressive and sudden deterioration and hazard occurrence prediction, among others. Therefore, further efforts are required for the accumulation of reliable and relevant data to increase the level of confidence in the solutions provided by application of life-cycle probabilistic analysis, performance prediction and optimum management of bridges.

- Improvements in probabilistic and physical models for evaluating and comparing the risks and benefits associated with various alternatives for maintaining or improving the reliability of existing bridges are needed (Frangopol 2011).

- Further and continued promotion of an integrated risk-based optimization approach as the rational basis for understanding the effect of technological, economical, environmental, and social interactions on the life-cycle performance and cost of bridges is required.
- Finally, encouraging education and research in the fields of risk acceptance and communication especially on educating stakeholders on their perception and tolerance of risk associated with bridges is necessary (Ellingwood 2005; Frangopol and Ellingwood 2010; Frangopol 2011; Cheng et al. 2020; Gong and Frangopol 2020; Liu et al. 2021). Also, developing tools and decision-making frameworks for engineers for cost-effectively incorporating lifetime resilience and sustainability taking into account climate change into the design of new bridges and the repair and rehabilitation of existing bridges is needed.

References

AASHTO (1992). Standard Specifications for Highway Bridges, 15th Edition. American Association of State Highway and Transportation Officials (AASHTO), Washington, D.C.

AASHTO (1994). LRFD Bridge Design Specifications, 1st Edition. American Association of State Highway and Transportation Officials (AASHTO), Washington, D.C.

AASHTO (1996). Standard Specifications for Highway Bridges, 16th Edition. American Association of State Highway and Transportation Officials (AASHTO), Washington, D.C.

AASHTO (2002). AASHTO Standard Specifications for Highway Bridges and Interim Specifications, 17th Edition. American Association of State Highway and Transportation Officials (AASHTO), Washington, D.C.

AASHTO (2007a). AASHTO Maintenance Manual for Roadways and Bridges, 4th Edition. American Association for State Highway and Transportation Officials (AASHTO), Washington, D.C.

AASHTO. (2007b). LRFD Bridge Design Specifications, 4th Edition. American Association of State Highway and Transportation Officials (AASHTO), Washington, D.C.

AASHTO (2010). Standard Specifications for Highway Bridges, 17th Edition. American Association of State Highway and Transportation Officials (AASHTO), Washington, D.C.

AASHTO (2011). AASHTO Guide Manual for Bridge Element Inspection, 1st Edition. American Association for State Highway and Transportation Officials (AASHTO), Washington, D.C.

AASHTO (2015). AASHTOW are Bridge Management User Manual BrM Version 5.2. American Association for State Highway and Transportation Officials (AASHTO), Washington, D.C.

AASHTO (2017a). LRFD Bridge Design Specification, 8th Edition. American Association of State Highway and Transportation Officials (AASHTO), Washington, D.C.

AASHTO (2017b). The Manual for Bridge Evaluation, 3rd Edition. American Association of State Highway and Transportation Officials (AASHTO), Washington, D.C.

Adey, B., Hajdin, R. and Brühwiler, E. (2003). Supply and demand system approach to development of bridge management strategies. *Journal of Infrastructure Systems*, ASCE, (9)3: 117–131.

Agrawal, A.K., Khan, M.A., Yi, Z. and Aboobaker, N. (2007). Handbook of Scour Countermeasures Designs. New Jersey Dept. of Transportation, Trenton, NJ.

Akbar, M.A., Qidwai, U. and Jahanshahi, M.R. (2019). An evaluation of image-based structural health monitoring using integrated unmanned aerial vehicle platform. *Structural Control and Health Monitoring*, John Wiley & Sons Ltd, 26(1): e2276.

Akgül, F. (2002). Lifetime system reliability prediction of multiple structure types in a bridge network. Ph.D. Thesis, Department of Civil, Environmental and Architectural Engineering, University of Colorado, Boulder, CO.

Akgül, F. and Frangopol, D.M. (2004a). Bridge rating and reliability correlation: comprehensive study for different bridge types. *Journal of Structural Engineering*, ASCE, 130(7): 1063–1074.

Akgül, F. and Frangopol, D.M. (2004b). Lifetime performance analysis of existing steel girder bridge superstructures. *Journal of Structural Engineering*, ASCE, 130(12): 1875–1888.

Akgül, F. and Frangopol, D.M. (2004c). Lifetime performance analysis of existing prestressed concrete bridge superstructures. *Journal of Structural Engineering*, ASCE, 130(12): 1889–1903.

Akgül, F. and Frangopol, D.M. (2004d). Computational platform for predicting lifetime system reliability profiles for different structure types in a network. *Journal of Computing in Civil Engineering*, ASCE, 18(2): 92–104.

Akgül, F. and Frangopol, D.M. (2004e). Time-dependent interaction between load rating and reliability of deteriorating bridges. *Engineering Structures*, Elsevier, 26(12): 1751–1765.

Akgül, F. and Frangopol, D.M. (2005a). Lifetime performance analysis of existing reinforced concrete bridges. I: Theory. *Journal of Infrastructure Systems,* ASCE, 11(2): 122–128.

Akgül, F. and Frangopol, D.M. (2005b). Lifetime performance analysis of existing reinforced concrete bridges. II: Application. *Journal of Infrastructure Systems,* ASCE, 11(2): 129–141.

Akiyama, M., Frangopol, D.M. and Yoshida, I. (2010). Time-dependent reliability analysis of existing RC structures in a marine environment using hazard associated with airborne chlorides. *Engineering Structures*, Elsevier, 32(11): 3768–3779.

Akiyama, M., Frangopol, D.M. and Matsuzaki, H. (2011). Life-cycle reliability of RC bridge piers under seismic and airborne chloride hazards. *Earthquake Engineering & Structural Dynamics*, John & Wiley, 40(15): 1671–1687.

Akiyama, M., Frangopol, D.M., Arai, M. and Koshimura, S. (2013). Reliability of bridges under tsunami hazards: Emphasis on the 2011 Tohoku-oki earthquake. *Earthquake Spectra*, SAGE, 29: S295–S314.

Akiyama, M., Frangopol, D.M. and Mizuno, K. (2014). Performance analysis of Tohoku-Shinkasen viaducts affected by the 2011 Great East Japan earthquake. *Structure and Infrastructure Engineering*, Taylor & Francis, 10(9): 1228–1247.

Akiyama, M., Frangopol, D.M. and Ishibashi, H. (2020). Toward life-cycle reliability-, risk- and resilience-based design and assessment of bridges and bridge networks under independent and interacting hazards: emphasis on earthquake, tsunami and corrosion. *Structure and Infrastructure Engineering*, Taylor & Francis, 16(1): 26–50.

Alampalli, S. (2014). Bridge maintenance. Chapter 11. pp. 269–300. *In*: W-F. Chen and L. Duan (eds). Bridge Engineering Handbook, 2nd Edition, Vol. 5: Construction and Maintenance. CRC Press/Taylor & Francis Group, Boca Raton, London, New York.

Alencar, G., de Jesus, A., da Silva, J.G.S. and Calçada, R. (2019). Fatigue cracking of welded railway bridges: a review. *Engineering Failure Analysis*, Elsevier, 104: 154–176.

Al-Harthy, A.S. and Frangopol, D.M. (1994). Reliability-based design of prestressed concrete beams. *Journal of Structural Engineering*, ASCE, 120(11): 3156–3177.

Alipour, A. and Shafei, B. (2016). Seismic resilience of transportation networks with deteriorating components. *Journal of Structural Engineering*, ASCE, 142: C4015015.

Ang, A. H.-S. and Tang, W.H. (1984). Probability Concepts in Engineering Planning and Design: Decision, Risk and Reliability. Vol. 2, John Wiley & Sons, New York.

Ang, A.H.-S. (2011). Life-cycle considerations in risk-informed decisions for design of civil infrastructures. *Structure and Infrastructure Engineering*, Taylor & Francis, 7(1–2): 3–9.

Ang, A.H.-S. and Tang, W.H. (2007). Probability Concepts in Engineering: Emphasis on Applications to Civil and Environmental Engineering. 2nd Edition. New York, Wiley.

Arora, J.S. (2017). Introduction to Optimum Design. 4th Edition. Elsevier, UK.

Arora, P., Popov, B.N., Haran, B., Ramasubramanian, M., Popova, S. and White, R.E. (1997). Corrosion initiation time of steel reinforcement in a chloride environment—A one dimensional solution. *Corrosion Science*, Elsevier, 39(4): 739–759.

ASCE (2017a). Policy statement 418 – The role of the civil engineer in sustainable development. Committee for Sustainability, American Society of Civil Engineers, Reston, Virginia.

ASCE. (2017b). Report Card for America's Infrastructure. American Society of Civil Engineers, Reston, VA.

ASTM (2012). Standard Practice for Liquid Penetrant Examination for General Industry. E165–12, West Conshohocken, PA.

Ataei, N. and Padgett, J.E. (2013). Probabilistic modeling of bridge deck unseating during hurricane events. *Journal of Bridge Engineering*, ASCE, 18(4): 275–286.

Augusti, G., Ciampoli, M. and Frangopol, D.M. (1998). Optimal planning of retrofitting interventions on bridges in a highway network. *Engineering Structures*, Elsevier, 20(11): 933–939.

Bai, Q., Labi, S., Sinha, K.C. and Thompson, P.D. (2013). Multiobjective optimization for project selection in network-level bridge management incorporating decision-maker's preference using the concept of holism. *Journal of Bridge Engineering*, ASCE, 18(9): 879–889.

Baker, J.W., Schubert, M. and Faber, M.H. (2008). On the assessment of robustness. *Structural Safety*. Elsevier, 30: 253–267.

Banerjee, S., Vishwanath, B.S. and Devendiran, D.K. (2019). Multihazard resilience of highway bridges and bridge networks: a review. *Structure and Infrastructure Engineering*, Taylor & Francis, 15(12): 1694–1714.

Barone, G. and Frangopol, D.M. (2013). Hazard-based optimum lifetime inspection and repair planning for deteriorating structures. *Journal of Structural Engineering*, ASCE, 139(12): 4013017.

Barone, G. and Frangopol, D.M. (2014a). Reliability, risk and lifetime distributions as performance indicators for life-cycle maintenance of deteriorating structures. *Reliability Engineering & System Safety*, Elsevier, 123: 21–37.

Barone, G. and Frangopol, D.M. (2014b). Life-cycle maintenance of deteriorating structures by multi-objective optimization involving reliability, risk, availability, hazard and cost. *Structural Safety*, Elsevier, 48: 40–50.

Barone, G., Frangopol, D.M. and Soliman, M. (2014). Optimization of life-cycle maintenance of deteriorating structures considering expected annual system failure rate and expected cumulative cost. *Journal of Structural Engineering*, ASCE, 140(2): 04013043.

Bastidas-Arteaga, E. and Stewart, M.G. (2015). Damage risks and economic assessment of climate adaptation strategies for design of new concrete structures subject to chloride-induced corrosion. *Structural Safety*, Elsevier, 52: 40–53.

Bastidas-Arteaga, E., Chateauneuf, A., Sánchez-Silva, M., Bressolette, P.H. and Schoefs, F. (2010). Influence of weather and global warming in chloride ingress into concrete: a stochastic approach. *Structural Safety*, Elsevier, 32(4): 238–249.

Bastidas-Arteaga, E., Schoefs, F., Stewart, M.G. and Wang, X. (2013). Influence of global warming on durability of corroding RC structures: A probabilistic approach. *Engineering Structures*, Elsevier, 51: 259–266

Beck, J.L. and Katafygiotis, L.S. (1998). Updating models and their uncertainties. I: Bayesian statistical framework. *Journal of Engineering Mechanics*, ASCE, 124(4): 455–461.

Bell, M.G.H. and Iida, Y. (1997). Transportation Network Analysis. Wiley, Chichester, U.K.

Benayoun, R., Roy, B. and Sussman, B. (1966). ELECTRE: Une méthode pour guider le choix en présence de points de vue multiples. Note Travail, 49, SEMA-METRA International, Direction Scientifique.

Berens, A.P. (1989). NDE reliability analysis. pp. 689–701. *In*: Metal handbook, 9th Edition, Vol. 17. ASM International, Material Park, Ohio.

Berens, A.P. and Hovey, P.W. (1981). Evaluation of NDE reliability characterization. Air Force Wright-Aeronautical Laboratory, Wright-Patterson Air Force Base, Dayton, Ohio.

Biondini, F. and Frangopol, D.M. (2009). Lifetime reliability-based optimization of reinforced concrete cross-sections under corrosion. *Structural Safety*, Elsevier, 31(6): 483–489.

Biondini, F. and Frangopol, D.M. (2015). Design, assessment, monitoring and maintenance of bridges and infrastructure networks. *Structure and Infrastructure Engineering*, Taylor & Francis, 11(4): 413–414.

Biondini, F. and Frangopol, D.M. (2016). Life-cycle performance of deteriorating structural systems under uncertainty: review. *Journal of Structural Engineering*, ASCE, 142(9): F4016001.

Biondini, F. and Marchiondelli, A. (2008). Evolutionary design of structural systems with time-variant performance. *Structure and Infrastructure Engineering*, Taylor & Francis, 4(2): 163–176.

Biondini, F., Camnasio, E. and Titi, A. (2015). Seismic resilience of concrete structure under corrosion. *Earthquake Engineering & Structural Dynamics*, John & Wiley, 44: 2445–2466.

Birnbaum, Z.W. (1969). On the importance of different components in a multicomponent system. pp. 581–592. *In*: P.R. Krishnaiah (ed.). Multivariate Analysis-II. Academic Press, New York.

Bjerager, P. (1996). The program system PROBAN. pp. 347–360. *In*: O. Ditlevsen and H.O. Madsen (eds). Structural Reliability Methods. Wiley–Blackwell, Chichester.

Bocchini, P. and Frangopol, D.M. (2010). On the applicability of random field theory to transportation analysis. Proceedings of the 5th international conference on bridge maintenance, safety, and management (IABMAS2010), July 11–15, 2010, Philadelphia.

Bocchini, P. and Frangopol, D.M. (2011a). A stochastic computational framework for the joint transportation network fragility analysis and traffic flow distribution under extreme events. *Probabilistic Engineering Mechanics*, Elsevier, 26(2): 182–193.

Bocchini, P. and Frangopol, D.M. (2011b). A probabilistic computational framework for bridge network optimal maintenance scheduling. *Reliability Engineering & System Safety*, Elsevier, 96(2): 332–349.

Bocchini, P. and Frangopol, D.M. (2011c). Generalized bridge network performance analysis with correlation and time-variant reliability. *Structural Safety*, Elsevier, 33(2): 155–164.

Bocchini, P. and Frangopol, D.M. (2012a). Optimal resilience- and cost-based postdisaster intervention prioritization for bridges along a highway segment. *Journal of Bridge Engineering*, ASCE, 17(1): 117–129.

Bocchini, P. and Frangopol, D.M. (2012b). Restoration of bridge networks after an earthquake: multicriteria intervention optimization. *Earthquake Spectra*. SAGE, 28(2): 427–455.

Bocchini, P. and Frangopol, D.M. (2013). Connectivity-based optimal scheduling for maintenance of bridge networks. *Journal of Engineering Mechanics*, ASCE, 139(6): 760–769.

Bocchini, P., Decò, A. and Frangopol, D.M. (2012). Probabilistic functionality recovery model for resilience analysis. Proceedings of the Sixth International Conference on Bridge Maintenance, Safety and Management, (IABMAS2012), July 8–12, 2012, Stresa, Lake Maggiore, Italy.

Bocchini, P., Frangopol, D.M., Ummenhofer, T. and Zinke, T. (2014). Resilience and sustainability of the civil infrastructure: towards a unified approach. *Journal of Infrastructure Systems*, ASCE, 20(2): 04014004.

Bourinet, J.-M., Mattrand, C. and Dubourg, V. (2009). A review of recent features and improvements added to FERUM software. In Proceeding of the 10th International Conference on Structural Safety and Reliability (ICOSSAR'09), Osaka, Japan, 2009.

Briaud, J.L., Ting, F.C.K., Chen, H.C., Gudavalli, R., Perugu, S. and Wei, G. (1999). SRICOS: Prediction of scour rate in cohesive soils at bridge piers. *Journal of Geotechnical and Geoenvironmental Engineering*, ASCE, 125(4): 237–246.

Brockhoff, D. and Zitzler, E. (2006). Dimensionality reduction in multiobjective optimization with (partial) dominance structure preservation: Generalized minimum objective subset problems. TIK Report 247, ETH Zurich, Switzerland.

Brockhoff, D. and Zitzler, E. (2007). Improving hypervolume-based multiobjective evolutionary algorithms by using objective reduction methods. 2007 IEEE Congress on Evolutionary Computation, IEEE Press, Singapore. pp. 2086–2093.

Brockhoff, D. and Zitzler, E. (2009). Objective reduction in evolutionary multiobjective optimization: Theory and applications. *Evolutionary Computation,* MIT Press, 17(2): 135–166.

Brownjohn, J.M.W. (2007). Structural health monitoring of civil infrastructure. *Philosophical Transactions of the Royal Society A*, Royal Society Publishing, 365(1851): 589–622.

Brownjohn, J.M.W., Stefano, A.D., Xu, Y.-L., Wenzel, H. and Aktan, A.E. (2011). Vibration-based monitoring of civil infrastructure: challenges and successes. *Journal of Civil Structural Health Monitoring*, Springer, 1(3): 79–95.

Brundtland, H. (1987). Our Common Future, Oxford University Press, New York.

Bruneau, M. and Reinhorn, A.M. (2007). Exploring the concept of seismic resilience for acute care facilities. *Earthquake Spectra*, SAGE, 28(1): 41–62.

Bruneau, M., Chang, S.E., Eguchi, R.T., Lee, G.C., O'Rourke, T.D., Reinhorn, A.M., Shinozuka, M., Tierney, K., Wallace, W.A., von Winterfeldt, D. (2003). A Framework to quantitatively assess and enhance the seismic resilience of communities. *Earthquake Spectra*, SAGE, 19(4): 733–752.

Caltrans Seismic Design Criteria (CSDC) (2004). Caltrans Seismic Design Criteria. California Department of Transportation, Sacramento, CA.

Cambridge Systematics, Inc. (2009). Pontis Release 4.5 User Manual, AASHTO, Washington, D.C.

Campbell, L.E., Connor, R.J., Whitehead, J.M. and Washer, G.A. (2020). Benchmark for evaluating performance in visual inspection of fatigue cracking in steel bridges. *Journal of Bridge Engineering*, ASCE, 25(1): 4019128.

Carden, E.P. and Fanning, P. (2004). Vibration based condition monitoring: a review. *Structural Health Monitoring*, SAGE, 3(4): 355–377.

Casciati, F. and Fuggini, C. (2009). Engineering vibration monitoring by GPS: long duration records. *Earthquake Engineering and Engineering Vibration*, Springer, 8(3): 459–467.

Catbas, F.N., Susoy, M. and Frangopol, D.M. (2008). Structural health monitoring and reliability estimation: Long span truss bridge application with environmental monitoring data. *Engineering Structures*, Elsevier, 30(9): 2347–2359.

CDOT (1998). Pontis Bridge Inspection Coding Guide, Denver, Colorado.

CDOT (2018). Quality Assurance Procedure – Liquid Penetrant Inspection Procedure. QAP 5945, Colorado Department of Transportation (CDOT), CO.

Cheng, M., Yang, D.Y. and Frangopol, D.M. (2020). Investigation of the effects of time preference and risk perception on life-cycle management of civil infrastructure. *ASCE-ASME Journal of Risk and Uncertainty in Engineering Systems, Part A: Civil Engineering*, ASCE, 6(1): 04020001, 1–11.

Chib, S. and Greenberg, E. (1995). Understanding the Metropolis-Hastings algorithm. *The American Statistician*, Taylor & Francis, 49(4): 327–335.

Choe, D.E., Gardoni, P., Rosowsky, D. and Haukaas, T. (2009). Seismic fragility estimates for reinforced concrete bridges subject to corrosion. *Structural Safety*, Elsevier, 31(4): 275–283.

Chung, H.-Y., Manuel, L. and Frank, K.H. (2006). Optimal inspection scheduling of steel bridges using nondestructive testing techniques. *Journal of Bridge Engineering*, ASCE, 11(3): 305–319.

Cimellaro, G.P., Reinhorn, A.M. and Bruneau, M. (2010). Framework for analytical quantification of disaster resilience. *Engineering Structures*, Elsevier, 32(11): 3639–3649.

Clemena, G.G., Lozev, M.G., Duke Jr, J.C. and Sison Jr, M.F. (1995). Interim Report: Acoustic Emission Monitoring of Steel Bridge Members. FHWA/VTRC 95-IR1, Virginia Department of Transportation (VDOT), VA.

Coello Coello C.A. (2003). Evolutionary multi-objective optimization: a critical review. pp. 117–146. *In*: R. Sarker, M. Mohammadian and X. Yao (eds). Evolutionary Optimization. International Series in Operations Research & Management Science, Vol. 48. Springer, Boston, MA.

Cohn, M.Z. (1980). Nonlinear Design of Concrete Structures: Problems and Prospects, University of Waterloo Press, Ontario, Canada.

Coles, S. (2001). An Introduction to Statistical Modeling of Extreme Values. Springer-Verlag.

Connor, R.J. and McCarthy, R.J. (2006). Report on field measurements and uncontrolled load testing of the Lehigh River Bridge (SR-33). Phase II. Final report. ATLSS Rep. No. 06-12, Lehigh Univ., Bethlehem, PA.

Connor, R.J. and Santosuosso, J. B. (2002). Report on Field Measurements and Controlled Load Testing of the Lehigh River Bridge (SR-33). Final report. ATLSS Rep. No. 02–07, Lehigh Univ., Bethlehem, PA.

Connor, R.J., Hodgson, I.C., Mahmoud, H.N., Bowman, C.A. (2005). Field testing and fatigue evaluation of the I-79 Neville Island Bridge over the Ohio River. ATLSS Rep. No. 05–02, Lehigh Univ., Bethlehem, PA.

Copelan, J.E. (2014). Bridge inspection. Chapter 13. pp. 337–350. *In*: W-F. Chen and L. Duan (eds). Bridge Engineering Handbook, 2nd Edition, Vol. 5: Construction and Maintenance. CRC Press/Taylor & Francis Group, Boca Raton, London, New York.

Crank, J. (1975). The Mathematics of Diffusion, 2nd Edition. Oxford, Oxford University Press.

Crawshaw, J. and Chambers, J. (1984). A Concise Course in A-Level Statistics. Stanley Thornes (Publishers) Ltd.

Cui, Y., Geng, Z., Zhu, Q., Han, Y. (2017). Review: Multi-objective optimization methods and application in energy saving. *Energy*, Elsevier, 125(15): 681–704.

Daniels, P.E., Ellis, D.R. and Stockton, W.R. (1999). Techniques for Manual Estimation Road User Costs Associated with Construction Projects. Texas Transportation Institute, Arlington, Texas.

Darmawan, M.S. and Stewart, M.G. (2007). Spatial time-dependent reliability analysis of corroding pretensioned prestressed concrete bridge girders. *Structural Safety*, Elsevier, 29(1): 16–31.

Deb, K. (2001). Multi-Objective Optimization Using Evolutionary Algorithms. John Wiley & Sons, New York.

Deb, K. and Saxena, D. (2006). Searching for Pareto-optimal solutions through dimensionality reduction for certain large-dimensional multi-objective optimization problems. Proceedings of the IEEE Congress on Evolutionary Computation (CEC2006), July 16–21, 2006, Vancouver, Canada.

Decò, A., P. Bocchini, and D.M. Frangopol. (2013). A probabilistic approach for the prediction of seismic resilience of bridges. *Earthquake Engineering and Structural Dynamics*, Wiley, 42(10): 1469–1487.

Deng, H., Yeh, C.-H. and Willis, R.J. (2000). Inter-company comparison using modified TOPSIS with objective weights. *Computers & Operations Research*, Pergamon, 27(10): 963–973.

Denton, S. (2002). Data Estimates for Different Maintenance Options for Reinforced Concrete Cross Heads (Personal Communication), Parsons Brinckerhoff Ltd., Bristol, UK.

Der Kiureghian, A. and Ditlevsen, O. (2009). Aleatoy or epistemic? Does it matter? *Structural Safety*, Elsevier, 31(2): 105–112.

Der Kiureghian, A., Haukaas, T. and Fujimura, K. (2006). Structural reliability software at the University of California, Berkeley. *Structural Safety*, Elsevier, 28(1–2): 44–67.

DeWolf, J.T., Lauzon, R.G. and Culmo, M.P. (2002). Monitoring bridge performance. *Structural Health Monitoring*, SAGE, 1(2): 129–138.

Diakoulaki, D., Mavrotas, G. and Papayannakis, L. (1995). Determining objective weights in multiple criteria problems: the critic method. *Computers & Operations Research*, Pergamon, 22(7): 763–770.

Dilek, U. (2007). Ultrasonic pulse velocity in nondestructive evaluation of low quality and damaged concrete and masonry construction. *Journal of Performance of Constructed Facilities*, ASCE, 21(5): 337–344.

Doebling, S.W., Hemez, F.M., Peterson, L.D. and Farhat, C. (1997). Improved damage location accuracy using strain energy-based on mode selection criteria. *AIAA Journal*, 35(4): 693–699.

Doebling, S., Farrar, C. and Prime, M. (1998). A summary review of vibration-based damage identification methods. *The Shock and Vibration Digest*, 30: 91–105.

Dong, Y. and Frangopol, D.M. (2015). Risk and resilience assessment of bridges under mainshock and aftershocks incorporating uncertainties. *Engineering Structures*, Elsevier, 83: 198–208.

Dong, Y. and Frangopol, D.M. (2016a). Incorporation of risk and updating in inspection of fatigue-sensitive details of ship structures. *International Journal of Fatigue*, Elsevier, 82(3): 676–688.

Dong, Y. and Frangopol, D.M. (2016b). Probabilistic time-dependent multihazard life-cycle assessment and resilience of bridges considering climate change. *Journal of Performance of Constructed Facilities*, ASCE, 30(5): 4016034.

Dong, Y. and Frangopol, D.M. (2017). Adaptation optimization of residential buildings under hurricane threat considering climate change in a lifecycle context. *Journal of Performance of Constructed Facilities*, ASCE, 31(6): 04017099.

Dong, Y. and Frangopol, D.M. (2019). Life-cycle performance of infrastructure networks. Chapter 3. pp. 65–94. *In*: F. Biondini and D.M. Frangopol (eds). Life-Cycle Design, Assessment and Maintenance of Structures and Infrastructure Systems, ASCE, Reston, VA.

Dong, Y., Frangopol, D.M. and Saydam, D. (2013). Time-variant sustainability assessment of seismically vulnerable bridges subjected to multiple hazards. *Earthquake Engineering & Structural Dynamics*, John Wiley & Sons, 42(10): 1451–1467.

Dong, Y., Frangopol, D.M. and Saydam, D. (2014a). Sustainability of highway bridge networks under seismic hazard. *Journal of Earthquake Engineering*, Taylor & Francis, 18(1): 41–66.

Dong, Y., Frangopol, D.M. and Saydam, D. (2014b). Pre-earthquake multi-objective probabilistic retrofit optimization of bridge networks based on sustainability. *Journal of Bridge Engineering*, ASCE, 19(6): 4014018.

Dong, Y., Frangopol, D.M. and Sabatino, S. (2015). Optimizing bridge network retrofit planning based on cost-benefit evaluation and multi-attribute utility associated with sustainability. *Earthquake Spectra*, EERI, 31(4): 2255–2280.

DuraCrete (2000). Statistical Quantification of the Variables in the Limit State Functions: DuraCrete – Probabilistic Performance-Based Durability Design of Concrete Structures. EU-Brite EuRam III. Project BE95-1347/R9.

Ellenberg, A., Branco, L., Krick, A., Bartoli, I. and Kontsos, A. (2015). Use of unmanned aerial vehicle for quantitative infrastructure evaluation. *Journal of Infrastructure Systems*, ASCE, 21(3): 04014054, 1–8.

Ellingwood, B.R., (2005). Risk-informed condition assessment of civil infrastructure: State of practice and research issues. *Structure and Infrastructure Engineering*, Taylor & Francis, 1(1): 7–18.

Ellingwood, B.R. and Kinali, K. (2009). Quantifying and communicating uncertainty in seismic risk assessment. *Structural Safety*, Elsevier, 31(2): 179–187.

Embrechts, P., Klüppelberg, C. and Mikosch, T. (1997). *Modelling Extremal Events for Insurance and Finance*. Berlin: Springer Verlag.

Enevoldsen, I. and Sørensen, J. (1994). Reliability-based optimization in structural engineering. *Structural Safety*, Elsevier, 15(3): 169–196.

Enright, M.P. (1998). Time-Variant Reliability of Reinforced Concrete Bridges Under Environmental Attack. PhD thesis, Dept. of Civil, Environmental, and Architectural Engineering, University of Colorado, Boulder, CO.

Enright, M.P. and Frangopol, D.M. (1998a). Probabilistic analysis of resistance degradation of reinforced concrete bridge beams under corrosion. *Engineering Structures*, Elsevier, 20(11): 960–971.

Enright, M.P. and Frangopol, D.M. (1998b). Service-life prediction of deteriorating concrete bridges. *Journal of Structural Engineering*, ASCE, 124(3): 309–317.

Enright, M.P. and Frangopol, D.M. (1998c). Failure time prediction of deteriorating fail-safe structures. *Journal of Structural Engineering*, ASCE, 124(12): 1448–1457.

Enright, M.P. and Frangopol, D.M. (1999a). Condition prediction of deteriorating concrete bridges using Bayesian updating. *Journal of Structural Engineering*, ASCE, 125(10): 1118–1125.

Enright, M.P. and Frangopol, D.M. (1999b). Maintenance planning for deteriorating concrete bridges. *Journal of Structural Engineering*, ASCE, 125(12): 1407–1414.

Enright, M.P. and Frangopol, D.M. (1999c). Reliability-based condition assessment of deteriorating concrete bridges considering load redistribution. *Structural Safety*, Elsevier 21(2): 159–195.

Enright, M.P. and Frangopol, D.M. (2000). RELTSYS: a computer program for life prediction of deteriorating systems. *Structural Engineering and Mechanics*, Techno–Press, 9(6): 557–568.

Ericson, C. (2015). Hazard Analysis Techniques for System Safety. 2nd Edition. Wiley, New York.

Estes, A.C. and Frangopol, D.M. (1998). RELSYS: a computer program for structural system reliability analysis. *Structural Engineering Mechanics*, Techno Press, 6(8): 901–919.

Estes, A.C. and Frangopol, D.M. (1999). Repair optimization of highway bridges using system reliability approach. *Journal of Structural Engineering*, ASCE, 125(7): 766–775.

Estes, A.C. and Frangopol, D.M. (2001). Minimum expected cost-oriented optimal maintenance planning for deteriorating structures: application to concrete bridge decks. *Reliability Engineering & System Safety*, Elsevier, 73(3): 281–291.

Estes, A.C. and Frangopol. D.M. (2003). Updating bridge reliability based on bridge management systems visual inspection results. *Journal of Bridge Engineering,* ASCE, 8(6): 374–382.

Estes, A.C. and Frangopol, D.M. (2005). Life-cycle evaluation and condition assessment of structures. Chapter 36. pp. 36–1 to 36–51. *In*: W-F. Chen and E.M. Lui (eds). Structural Engineering Handbook, 2nd Edition. CRC Press.

Estes, A.C., Frangopol, D.M. and Foltz, S.D. (2004). Updating reliability of steel miter gates on locks and dams using visual inspection results. *Engineering Structures*, Elsevier, 26(3): 319–333.

Farrar, C.R. and Worden, K. (2007). An introduction to structural health monitoring. *Philosophical Transactions of the Royal Society A*, Royal Society Publishing, 365(1851): 303–315.

Fei, L., Xia, J., Feng, Y. and Liu, L. (2019). An ELECTRE-based multiple criteria decision making method for supplier selection using Dempster-Shafer theory. *IEEE Access*, IEEE, 7: 84701–84716.

FEMA (2009). HAZUS-MH MR4 Earthquake Model User Manual. Department of Homeland Security, Federal Emergency Management Agency (FEMA), Washington, D.C.

FEMA (2020). HAZUS Earthquake Model Technical Manual. Department of Homeland Security, Federal Emergency Management Agency (FEMA), Washington, D.C.

Feng, D. and Feng, M.Q. (2015). Vision-based multipoint displacement measurement for structural health monitoring. *Structural Control and Health Monitoring*, John Wiley & Sons, Ltd, 23(5): 876–890.

Feng, D. and Feng, M.Q. (2018). Computer vision for SHM of civil infrastructure: From dynamic response measurement to damage detection – a review. *Engineering Structures*, Elsevier, 156: 105–117.

FHWA (1995). Recording and Coding Guide for Structure Inventory and Appraisal of the Nation's Bridge, Report No. FHWA-PD 96-001, U.S. Department of Transportation, Washington, D.C.

FHWA (2001). Reliability of Visual Inspection for Highway Bridges, Volume 1: Final Report. Publication No. FHWA-RD-01-020, Federal Highway Administration (FHWA), Washington, D.C.

FHWA (2004). National Bridge Inspection Standards. Report No: 23 CFR Part 650 Subpart C., Federal Highway Administration (FHWA), Washington, D.C.

FHWA (2005). Bridge Preservation and Maintenance in Europe and South Africa. Report No. FHWA-PL-05-002, Federal Highway Administration (FHWA), Washington, D.C.

FHWA (2010). Underwater Bridge Repair, Rehabilitation, and Countermeasures. Publication No. FHWA-NHI-10-029, Federal Highway Administration, Washington, D.C.

FHWA (2012). Bridge Inspector's Reference Manual. Publication No. FHWA-NHI12-049, Federal Highway Administration (FHWA), Washington, D.C.

FHWA (2013a). LTBP Bridge Performance Primer, Report No. FHWA-HRT-13-051, U.S. Department of Transportation, Washington, D.C.

FHWA (2013b). Manual for Repair and Retrofit of Fatigue Cracks in Steel Bridges. FHWA Publication No. FHWA-IF-13-020, Arlington, VA, USA.

FHWA (2015). Bridge Maintenance Reference Manual. Publication No. FHWA-NHI-14-050, Federal Highway Administration (FHWA), Washington, D.C.

FHWA (2018). Bridge Preservation Guide: Maintaining a Resilient Infrastructure to Preserve Mobility. Federal Highway Administration (FHWA), Washington, D.C.

Fiorillo, G. and Ghosn, M. (2019). Risk-based importance factors for bridge networks under highway traffic loads. *Structure and Infrastructure Engineering*, Taylor & Francis, 15(1): 113–126.

Fisher, J.W. (1984). Fatigue and Fracture in Steel Bridges. Wiley, New York.

Fisher, J.W., Kulak, G.L. and Smith, I.F. (1998). A Fatigue Primer for Structural Engineers. National Steel Bridge Alliance, Chicago, IL, USA.

Flintsch, G. and Chen, C. (2004). Soft computing applications in infrastructure management. *Journal of Infrastructure Systems*, 10(4): 157–166.

Frangopol, D.M. (1985). Structural optimization using reliability concepts. *Journal of Structural Engineering*, ASCE, 111(11): 2288–2301.

Frangopol, D.M. (2011). Life-cycle performance, management, and optimization of structural systems under uncertainty: Accomplishments and challenges. *Structure and Infrastructure Engineering*, Taylor & Francis, 7(6): 389–413.

Frangopol, D.M. (2018). Structures and Infrastructure Systems: Life-Cycle Performance, Management, and Optimization. Routledge/Taylor & Francis Group.

Frangopol, D.M. and Bocchini, P. (2011). Resilience as optimization criterion for the rehabilitation of bridges belonging to a transportation network subject to earthquake. Proceedings of the ASCE 2011 Structures Congress SEI 2011, Las Vegas, NV.

Frangopol, D.M. and Bocchini, P. (2012). Bridge network performance, maintenance and optimisation under uncertainty: accomplishments and challenges. *Structure and Infrastructure Engineering*, Taylor & Francis, 8(4): 341–356.

Frangopol, D.M. and Curley, J.P. (1987). Effects of damage and redundancy on structural reliability. *Journal of Structural Engineering*, ASCE, 113(7): 1533–1549.

Frangopol, D.M. and Estes, A.C. (1997). Lifetime bridge maintenance strategies based on system reliability. *Structural Engineering International*, IABSE, 7(3): 193–198.

Frangopol, D.M. and Ellingwood, B.R. (2010). Life-cycle performance, safety, reliability and risk of structural systems: a framework for new challenges. Editorial. *Structure*. A joint publication of NCSEA/CASE/SEI, March 7, 2010.

Frangopol, D.M. and Klisinski, M. (1989). Material behavior and optimum design of structural systems. *Journal of Structural Engineering*, ASCE, 115(5): 1054–1075

Frangopol, D.M. and Kim, S. (2011). Service life, reliability and maintenance of civil structures. Chapter 5. pp. 145–178. *In*: L.S. Lee and V. Karbari (eds). Service Life Estimation and Extension of Civil Engineering Structures. Woodhead Publishing Ltd., Cambridge, U.K.

Frangopol, D.M. and Kim, S. (2014a). Bridge health monitoring. Chapter 10. pp. 247–268. *In*: W-F. Chen and L. Duan (eds). Bridge Engineering Handbook, 2nd Edition, Vol. 5: Construction and Maintenance. CRC Press/Taylor & Francis Group, Boca Raton, London, New York.

Frangopol, D.M. and Kim, S. (2014b). Life-cycle analysis and optimization. Chapter 18. pp. 537–566. *In*: W-F. Chen and L. Duan (eds). Bridge Engineering Handbook, 2nd Edition, Vol. 5: Construction and Maintenance. CRC Press/Taylor & Francis Group, Boca Raton, London, New York.

Frangopol, D.M. and Kim, S. (2014c). Prognosis and life-cycle assessment based on SHM information. Chapter 5. pp. 145–171. *In*: M.L. Wang, J. Lynch and H. Sohn (eds.) Part II. Data Interrogation and Decision Making in Sensor Technologies for Civil Infrastructures: Performance Assessment and Health Monitoring. Woodhead Publishing Ltd., Cambridge.

Frangopol, D.M. and Kim, S. (2019). Life-Cycle of Structures under Uncertainty: Emphasis on Fatigue-Sensitive Civil and Marine Structures. CRC Press, Boca Raton.

Frangopol, D.M. and Liu, M. (2007a). Bridge network maintenance optimization using stochastic dynamic programming. *Journal of Structural Engineering*, ASCE, 133(12): 1772–1782.

Frangopol, D.M. and Liu, M. (2007b). Maintenance and management of civil infrastructure based on condition, safety, optimization, and life-cycle cost. *Structure and Infrastructure Engineering*, Taylor & Francis, 3(1): 29–41.

Frangopol, D.M. and Maute, K. (2003). Life-cycle reliability-based optimization of civil and aerospace structures. *Computers & Structures*, Elsevier, 81(7): 397–410.

Frangopol, D.M. and Messervey, T.B. (2011). Effect of monitoring on reliability of structures. Chapter 18. pp. 515–560. *In*: B. Bakht, A.A. Mufti and L.D. Wegner (eds). Monitoring Technologies for Bridge Management. Multi-Science Publishing Co. Ltd., U.K.

Frangopol, D.M. and Nakib, R. (1991). Redundancy in highway bridges. *Engineering Journal*, American Institute of Steel Construction (AISC), Chicago, IL, 28(1): 45–50.

Frangopol, D.M. and Saydam, D. (2014). Structural performance indicators for bridges. Chapter 9. pp. 185–206. *In*: W-F. Chen and L. Duan (eds). Bridge Engineering Handbook, 2nd Edition, Vol. 5: Construction and Maintenance. CRC Press/Taylor & Francis Group, Boca Raton, London, New York.

Frangopol, D.M. and Soliman, M. (2016). Life-cycle of structural systems: recent achievements and future directions. *Structure and Infrastructure Engineering,* Taylor & Francis, 12(1): 1–20.

Frangopol, D.M., Dong, Y. and Sabatino, S. (2017). Bridge life-cycle performance and cost: Analysis, prediction, optimization, and decision-making. *Structure and Infrastructure Engineering,* Taylor & Francis, 13(10): 1239–1257.

Frangopol, D.M., Kong, J.S. and Gharaibeh, E.S. (2001). Reliability-based life-cycle management of highway bridges. *Journal of Computing in Civil Engineering,* ASCE, 15(1): 27–34.

Frangopol, D.M., Lin, K-Y. and Estes, A.C. (1997a). Reliability of reinforced concrete girders under corrosion attack. *Journal of Structural Engineering,* ASCE, 123(3): 286–297.

Frangopol, D.M., Lin, K-Y. and Estes, A.C. (1997b). Life-cycle cost design of deteriorating structures. *Journal of Structural Engineering,* ASCE, 123(10): 1390–1401.

Frangopol, D.M., Saydam, D. and Kim, S. (2012). Maintenance, management, life-cycle design and performance of structures and infrastructures: a brief review. *Structure and Infrastructure Engineering,* Taylor & Francis, 8(1): 1–25.

Frangopol, D.M., Strauss, A. and Kim, S. (2008a). Bridge reliability assessment based on monitoring. *Journal of Bridge Engineering,* ASCE, 13(3): 258–270.

Frangopol, D.M., Strauss, A. and Kim, S. (2008b). Use of monitoring extreme data for the performance prediction of structures: general approach. *Engineering Structures,* Elsevier, 30 (12): 3644–3653.

Fu, G. and Frangopol, D.M. (1990a). Reliability-based vector optimization of structural systems. *Journal of Structural Engineering,* ASCE, 116(8): 2143–2161.

Fu, G. and Frangopol, D.M. (1990b). Balancing weight, system reliability and redundancy in a multi-objective optimization framework. *Structural Safety,* Elsevier, 7(2–4): 165–175.

Fujino, Y. and Siringoringo, D. (2013). Vibration mechanisms and controls of long-span bridges: a review. *Structural Engineering International,* Taylor & Francis, 23(3): 248–268.

Fukuda, Y., Feng, M.Q. and Shinozuka, M. (2010). Cost-effective vision-based system for monitoring dynamic response of civil engineering structures. *Structural Control and Health Monitoring,* John Wiley & Sons, Ltd, 17(8): 918–936.

Furuta, H., Frangopol, D.M. and Nakatsu, K. (2011). Life-cycle cost optimization with emphasis on balancing structural performance and seismic risk of road network. *Structure and Infrastructure Engineering,* Taylor & Francis, 7(1–2): 65–74

Furuta, H., Kameda, T., Nakahara, K., Takahashi, Y. and Frangopol, D.M. (2006). Optimal bridge maintenance planning using improved multi-objective genetic algorithm. *Structure and Infrastructure Engineering,* Taylor & Francis, 2(1): 33–41.

Ghosn, M. and Moses, F. (1998). Redundancy in Highway Bridge Superstructures. National Cooperative Highway Research Program (NCHRP), Washington, D.C.

Ghosn, M., Moses, F. and Frangopol, D.M. (2010). Redundancy and robustness of highway bridge superstructures and substructures. *Structure and Infrastructure Engineering,* Taylor & Francis, 6(1–2): 257–278.

Ghosn, M., Dueñas-Osorio, L., Frangopol, D.M., McAllister, T.P., Bocchini, P., Manuel L., Ellingwood, B.R., Arangio, S., Bontempi, F., Shah, M., Akiyama, M., Biondini, F., Hernandez, S. and Tsiatas, G. (2016a). Performance indicators for structural systems and infrastructure networks. *Journal of Structural Engineering,* ASCE, 142(9): F4016003, 1–18.

Ghosn, M., Frangopol, D.M., McAllister, T.P., Shah, M., Diniz, S., Ellingwood, B.R., Manuel, L., Biondini, F., Catbas, N., Strauss, A. and Zhao, Z.L. (2016b). Reliability-based structural performance indicators for structural members. *Journal of Structural Engineering*, ASCE, 142(9): F4016002, 1–13.

Gilks, W.R., Richardson, S. and Spiegelhalter, D.J. (1996). Markov Chain Monte Carlo in Practice. Chapman & Hall, London.

Gong, C. and Frangopol, D.M. (2019). An efficient time-dependent reliability method. *Structural Safety*, Elsevier, 85: 101864, 1–7.

Gong, C. and Frangopol, D.M. (2020). Condition-based multi-objective maintenance decision-making for highway bridges considering risk perceptions. *Journal of Structural Engineering*, ASCE, 146(5): 04020051, 1–13.

Govindan, K. and Jepsen, M.B. (2016). ELECTRE: A comprehensive literature review on methodologies and applications. *European Journal of Operational Research*, Elsevier, 250(1): 1–29.

Grosse, C. U. (2010). Acoustic emission (AE) evaluation of reinforced concrete structures. Chapter 10. pp. 185–214. *In*: C. Maierhofer, H.-W. Reinhardt and G. Dobmann (eds). Non-Destructive Evaluation of Reinforced Concrete Structures. Woodhead Publishing Series in Civil and Structural Engineering. Woodhead Publishing.

Guedes Soares, C. and Garbatov, Y. (1996a). Fatigue reliability of the ship hull girder accounting for inspection and repair. *Reliability Engineering and System Safety*, Elsevier, 51(3): 341–351.

Guedes Soares, C. and Garbatov, Y. (1996b). Fatigue reliability of the ship hull girder. *Marine Structures*, Elsevier, 9(3–4): 495–516.

Gumbel, E.J. (1958). Statistics of Extremes. Columbia Univ. Press, New York, NY.

Gunantara, N. (2018). A review of multi-objective optimization: Methods and its applications. Cogent Engineering, Taylor & Francis, 5(1): 1502242.

Guo, T., Frangopol, D.M. and Chen, Y. (2012). Fatigue reliability assessment of steel bridge details integrating weigh-in-motion data and probabilistic finite element analysis. *Computers and Structures*, 112–113: 245–257.

Haardt, P. (2002). Development of a bridge management system for the German highway network. Proceedings of the First International Conference on Bridge Maintenance, Safety and Management (IABMAS 2002), July 14–17, 2002, Barcelona, Spain.

Haardt, P. and Holst, R. (2008). The German approach to bridge management current status and future development. Proceedings of the Tenth International Conference on Bridge and Structure Management, October 20–22, 2008, Buffalo, NewYork.

Haghani. R., Al-Emrani, M. and Heshmati, M. (2012). Fatigue-prone details in steel bridges. *Buildings*, MDPI, 2(4): 456–476.

Hajdin, R. (2004). KUBA-MS: The Swiss bridge management system. Proceedings of Structures 2001, May 21–23, 2001, Washington, D.C., U.S.

Hajdin, R. (2008). KUBA 4.0: The Swiss road structure management system. Proceedings of the Tenth International Conference on Bridge and Structure Management, October 20–22, 2008, Buffalo, NewYork.

Hallberg, D. and Racutanu, G. (2007). Development of the Swedish bridge management system by introducing a LMS concept. *Materials and Structures*, Springer, 40(6): 627–639.

Hartle, R.A., Amrhein, W.J., Wilson, K.E., III, Bauhman, D.R. and Tkacs, J.J. (1995). Bridge inspector's training manual 90. PD-91-015, Federal Highway Administration (FHWA), Washington, D.C.

Hashin, Z. and Rotem, A. (1978). A cumulative damage theory of fatigue failure. *Materials Science and Engineering*, Elsevier, 34(2): 147–160.

Hasting, W.K. (1970). Monte Carlo sampling methods using Markov chains and their applications. *Biometrika*, Oxford University Press, 57(1): 97–109.

Hatami-Marbini, A. and Tavana, M. (2011). An extension of the Electre I method for group decision-making under a fuzzy environment. *Omega*, Elsevier, 39: 373–386.

Hawk, H. and Small, E.P. (1998). The BRIDGIT Bridge Management System. *Structural Engineering International*, Taylor & Francis, 8(4): 309–314.

Henry, D. and Emmanuel Ramirez-Marquez, J. (2012). Generic metrics and quantitative approaches for system resilience as a function of time. *Reliability Engineering & System Safety*, Elsevier, 99: 114–122.

Herrman, J.W. (2015). Engineering Decision Making and Risk Management. Wiley, NJ.

Hessami, A.G. (1999). Risk management: a system paradigm. *System Engineering*, Jhon Wiley & Sons, 2(3): 156–167.

Holling, C. (1973). Resilience and stability of ecological systems. *Annual Review of Ecological Systems*, Annual Review, 4(1): 1–23.

Hu, X. and Madanat, S. (2015). Determination of optimal MR&R policies for retaining life-cycle connectivity of bridge networks. *Journal of Infrastructure Systems*, ASCE, 21(2): 4014042.

Hu, X., Daganzo, C. and Madanat, S. (2015). A reliability-based optimization scheme for maintenance management in large-scale bridge networks. *Transportation Research Part C: Emerging Technologies*, Elsevier, 55: 166–178.

Hurt, M. and Schrock, S.D. (2016). Highway Bridge Maintenance Planning and Scheduling. Butterworth-Heinemann, UK.

Hwang, C.-L., Lai, Y.-J. and Liu, T.-Y. (1993). A new approach for multiple objective decision making. *Computers & Operations Research*, Pergamon, 20(8): 889–899.

Hwang, W. and Han, K.S. (1986). Cumulative damage models and multi-stress fatigue life prediction. *Journal of Composite Materials*, SAGE, 20(2): 125–153.

IAEA (2002). Guidebook on Non-Destructive Testing of Concrete Structures. International Atomic Energy Agency (IAEA), Vienna, Austria.

Im, S.B., Hurlebaus, S. and Kang, Y.J. (2013). Summary review of GPS technology for structural health monitoring. *Journal of Structural Engineering*, ASCE, 139(10): 1653–1664.

IPCC (2014). Climate Change 2014: Synthesis Report. Intergovernmental Panel on Climate Change (IPCC), Geneva, Switzerland.

Irwin, G.R. (1958). The crack-extension-force for a crack at a free surface boundary. Report No. 5120, Naval Research Laboratory, Washington, D.C.,

Ishibashi, H., Akiyama, M., Frangopol, D.M., Koshimura, S., Kojima, T. and Nanami, K. (2021). Framework for estimating the risk and resilience of road networks with bridges and embankments under both seismic and tsunami hazards. *Structure and Infrastructure Engineering*, Taylor & Francis, 17(4): 494–514.

Jahanshahi, M.R., Kelly, J.S., Masri, S.F. and Sukhatme, G.S. (2009). A survey and evaluation of promising approaches for automatic image-based defect detection of bridge structures. *Structure and Infrastructure Engineering*, Taylor & Francis, 5(6): 455–486.

Jiménez, A., Ríos-Insua, S. and Mateos, A. (2003). A decision support system for multi-attribute utility evaluation based on imprecise assignments. *Decision Support Systems*, Elsevier, 36(1): 65–79.

Kabir, G., Sadiq, R. and Tesfamariam, S. (2014). A review of multi-criteria decision-making methods for infrastructure management. *Structure and Infrastructure Engineering*, 10(9): 1176–1210.

Kafali, C. and Grigoriu, M. (2005). Rehabilitation decision analysis. Proceeding of the 9th International Conference on Structural Safety and Reliability (ICOSSAR'05), June 19–23, 2005, Rome, Italy.

Kafandaris, S. (2002). ELECTRE and decision support: methods and applications in engineering and infrastructure investment. *Journal of the Operational Research Society*, Taylor & Francis, 53(12): 1396–1397.

Karandikar, J.M., Kim, N.H. and Schmitz, T.L. (2012). Prediction of remaining useful life for fatigue-damaged structures using Bayesian inference. *Engineering Fracture Mechanics*, Elsevier, 96: 588–605.

Kececioglu, D. (1995). Maintainability, Availability, & Operational Readiness Engineering. Prentice-Hall, NJ.

Kim, S. and Frangopol, D.M. (2009). Optimal decision making of structural health monitoring under uncertainty. Proceedings of the Tenth International Conference on Structural Safety and Reliability, ICOSSAR2009, Osaka, Japan, September 13–17, 2009.

Kim, S. and Frangopol, D.M. (2010). Optimal planning of structural performance monitoring based on reliability importance assessment. *Probabilistic Engineering Mechanics*, Elsevier, 25(1): 86–98.

Kim, S. and Frangopol, D.M. (2011a). Cost-effective lifetime structural health monitoring based on availability. *Journal of Structural Engineering*, ASCE, 137(1): 22–33.

Kim, S. and Frangopol, D.M. (2011b). Cost-based optimum scheduling of inspection and monitoring for fatigue-sensitive structures under uncertainty. *Journal of Structural Engineering*, ASCE, 137(11): 1319–1331.

Kim, S. and Frangopol, D.M. (2011c). Inspection and monitoring planning for RC structures based on minimization of expected damage detection delay. *Probabilistic Engineering Mechanics,* Elsevier, 26(2): 308–320.

Kim, S. and Frangopol, D.M. (2011d). Optimum inspection planning for minimizing fatigue damage detection delay of ship hull structures. *International Journal of Fatigue*, Elsevier, 33(3): 448–459.

Kim, S., Frangopol, D.M. and Zhu, B. (2011). Probabilistic optimum inspection/repair planning to extend lifetime of deteriorating structures. *Journal of Performance of Constructed Facilities*, ASCE, 25(6): 534–544.

Kim, S. and Frangopol, D.M. (2012). Probabilistic bicriterion optimum inspection/ monitoring planning: applications to naval ships and bridges under fatigue. *Structure and Infrastructure Engineering*, Taylor & Francis, 8(10): 912–927.

Kim, S., Frangopol, D.M. and Soliman, M. (2013). Generalized probabilistic framework for optimum inspection and maintenance planning. *Journal of Structural Engineering*, ASCE, 139(3): 435–447.

Kim, S. and Frangopol, D.M. (2017). Efficient multi-objective optimisation of probabilistic service life management. *Structure and Infrastructure Engineering*, Taylor & Francis, 13(1): 147–159.

Kim, S. and Frangopol, D.M. (2018a). Decision making for probabilistic fatigue inspection planning based on multi-objective optimization. *International Journal of Fatigue*, Elsevier, 111: 356–368.

Kim, S. and Frangopol, D.M. (2018b). Multi-objective probabilistic optimum monitoring planning considering fatigue damage detection, maintenance, reliability, service life and cost. *Structural and Multidisciplinary Optimization*, Springer, 57(1): 39–54.

Kim, S., Ge, B. and Frangopol, D.M. (2019). Effective optimum maintenance planning with updating based on inspection information for fatigue-sensitive structures. *Probabilistic Engineering Mechanics*, Elsevier, 58: 103003.

Kim, S. and Frangopol, D.M. (2020). Computational platform for probabilistic optimum monitoring planning for effective and efficient service life management. *Journal of Civil Structural Health Monitoring*, Springer, 10(1): 1–15.

Kim, S., Ge, B. and Frangopol, D.M. (2020). Optimum target reliability determination for efficient service life management of bridge networks. *Journal of Bridge Engineering*, ASCE, 25(10): 4020087: 1–14.

Kong, J.S. (2001). Lifetime maintenance strategies for deteriorating structures. Ph.D. Thesis, Department of Civil, Environmental and Architectural Engineering, University of Colorado, Boulder, CO.

Kong, J.S. and Frangopol, D.M. (2002). Life-cycle performance prediction of steel/concrete composite bridges. *International Journal of Steel Structures*, KSSC, 2(1): 13–19.

Kong, J.S. and Frangopol, D.M. (2003a). Evaluation of expected life-cycle maintenance cost of deteriorating structures. *Journal of Structural Engineering*, ASCE, 129(5): 682–691.

Kong, J.S. and Frangopol, D.M. (2003b). Life-cycle reliability-based maintenance cost optimization of deteriorating structures with emphasis on bridges. *Journal of Structural Engineering*, ASCE, 129(6): 818–828.

Kong, J. and Frangopol, D.M. (2004a). Prediction of reliability and cost profiles of deteriorating bridges under time- and performance-controlled maintenance. *Journal of Structural Engineering*, ASCE, 130(12): 1865–1874.

Kong, J. and Frangopol, D.M. (2004b). Cost–reliability interaction in life-cycle cost optimization of deteriorating structures. *Journal of Structural Engineering*, ASCE, 130(11): 1704–1712.

Kong, J. and Frangopol, D.M. (2005). Probabilistic optimization of aging structures considering maintenance and failure costs. *Journal of Structural Engineering*, ASCE, 131(4): 600–616.

Kwon, K. and Frangopol, D.M. (2010). Bridge fatigue reliability assessment using probability density functions based on field monitoring data. *International Journal of Fatigue*, Elsevier, 32(8): 1221–1232.

Kwon, K. and Frangopol, D.M. (2011). Bridge fatigue assessment and management using reliability-based crack growth and probability of detection models. *Probabilistic Engineering Mechanics*, Elsevier, 26(3): 471–480.

Kwon, K., Frangopol, D.M. and Kim, S. (2013). Fatigue performance assessment and service life prediction of high-speed ship structures based on probabilistic lifetime sea loads. *Structure and Infrastructure Engineering*, Taylor & Francis, 9(2): 102–115.

Kwon, K., Frangopol, D.M. and Soliman, M. (2012). Probabilistic fatigue life estimation of steel bridges based on a bi-linear S-N approach. *Journal of Bridge Engineering*, ASCE, 17(1): 58–70.

Leadbetter, M.R., Lindgren, G. and Rootzén, H. (1983). Extremes and Related Properties of Random Sequences and Processes. Springer-Verlag.

Lee, L.S., Karbhari, V.M. and Sikorsky, C. (2007). Structural health monitoring of CFRP strengthened bridge decks using ambient vibrations. *Structural Health Monitoring*, SAGE, 6(3): 199–214.

Lee, W.K. and Lin, C.Y. (2008). Developing a sustainability evaluation system in Taiwan to support infrastructure investment decisions. *International Journal of Sustainable Transportation*, Taylor & Francis, 2(3): 194–212.

Leemis, L.M. (1986). Lifetime distribution identities. *IEEE Transactions on Reliability*, IEEE, 35(2): 170–174.

Leemis, L.M. (2009). Reliability: Probabilistic Models and Statistical Methods. 2nd Edition. Ascended Ideas, U.S.

Li, Z.X., Chan, T.H.T. and Ko, J.M. (2002). Evaluation of typhoon induced fatigue damage for Tsing Ma Bridge. *Engineering Structures*, Elsevier, 24(8): 1035–1047.

Li, H., Frangopol, D.M., Soliman, M. and Xia, H. (2016). Fatigue reliability assessment of railway bridges based on probabilistic dynamic analysis of coupled train-bridge system. *Journal of Structural Engineering*, ASCE, 142(3): 04015158, 1–16.

Li, Y., Dong, Y., Frangopol, D.M. and Gautam, D. (2020). Resilience assessment of highway bridges under multiple natural hazards. *Structure and Infrastructure Engineering*, Taylor & Francis, 16(4): 626–641.

Lin, K-Y. (1995). Reliability-based minimum life cycle cost design of reinforced concrete girder bridges. Ph.D. Thesis, Department of Civil, Environmental and Architecture Engineering, University of Colorado, Boulder, CO.

Lin, K.-Y. and Frangopol, D.M. (1996). Reliability-based optimum design of reinforced concrete girders. *Structural Safety*, Elsevier, 18(2–3): 239–258.

Lind, N.C. (1995). A measure of vulnerability and damage tolerance. *Reliability Engineering & System Safety*, Elsevier, 43(1): 1–6.

Liu, L., Frangopol, D.M., Mondoro, A. and Yang, D.Y. (2018). Sustainability-Informed bridge ranking under scour based on transportation network performance and multiattribute utility. *Journal of Bridge Engineering*, ASCE, 23(10): 04018082, 1–12.

Liu, L., Yang, D.Y. and Frangopol, D.M. (2020). Network-level risk-based framework for optimal bridge adaptation management considering scour and climate change. *Journal of Infrastructure Systems*, ASCE, 26(1): 4019037.

Liu, L., Yang, D.Y. and Frangopol, D.M. (2021). Determining target reliability index of structures based on cost optimization and acceptance criteria for fatality risk. ASCE-ASME. *Journal of Risk and Uncertainty in Engineering Systems, Part A: Civil Engineering*, ASCE, 7(2): 0401013, 1–13.

Liu, M. and Frangopol, D.M. (2005a). Balancing connectivity of deteriorating bridge networks and long-term maintenance cost through optimization. *Journal of Bridge Engineering*, ASCE, 10(4): 468–481.

Liu, M. and Frangopol, D.M. (2005b). Multi-objective maintenance planning optimization for deteriorating bridges considering condition, safety and life-cycle cost. *Journal of Structural Engineering*, ASCE, 131(5): 833–842.

Liu, M. and Frangopol, D.M. (2005c). Time-dependent bridge network reliability: novel approach. *Journal of Structural Engineering*, ASCE, 131(2): 329–337.

Liu, M. and Frangopol, D.M. (2005d). Bridge annual maintenance prioritization under uncertainty by multi-objective combinatorial optimization. *Computer Aided Civil and Infrastructure Engineering*, Blackwell Publishing, 20(5): 343–353.

Liu, M. and Frangopol, D.M. (2006a). Probability-based bridge network performance evaluation. *Journal of Bridge Engineering*, ASCE, 11(5): 633–641.

Liu, M. and Frangopol, D.M. (2006b). Optimizing bridge network maintenance management under uncertainty with conflicting criteria: Life-cycle maintenance, failure, and user costs. *Journal of Structural Engineering*, ASCE, 132(11): 1835–1845.

Liu, M., Frangopol, D.M. and Kim, S. (2009a). Bridge safety evaluation based on monitored live load effects. *Journal of Bridge Engineering*, ASCE, 14(4): 257–269.

Liu, M., Frangopol, D.M. and Kim, S. (2009b). Bridge system performance assessment from structural health monitoring: a case study. *Journal of Structural Engineering*, ASCE, 135(6): 733–742.

Liu, M., Frangopol, D.M. and Kwon, K. (2010a). Fatigue reliability assessment of retrofitted steel bridges integrating monitoring data. *Structural Safety*, Elsevier, 32(1): 77–89.

Liu, M., Frangopol, D.M. and Kwon, K. (2010b). Optimization of retrofitting distortion-induced fatigue cracking of steel bridges using monitored data under uncertainty. *Engineering Structures*, Elsevier, 32(11): 3467–3477.

Liu, P.-L., Lin, H.-Z., Der Kiureghian, A. (1989). *CALREL User Manual*. Technical Report UCB/SEMM-89/18, Department of Civil Engineering, University of California, Berkeley, California.

Liu, Y. and Frangopol, D.M. (2019). Utility and information analysis for optimum inspection of fatigue-sensitive structures. *Journal of Structural Engineering*, ASCE, 145(2): 4018251.

Lounis, Z. and McAllister, T.P. (2016). Risk-based decision making for sustainable and resilient infrastructure systems. *Journal of Structural Engineering*, ASCE, 142(9): F4016005.

Mackie, P. and Nellthorp, J. (2001). Cost-benefit analysis in transport. Chapter 10. pp. 143–174. *In*: K.J. Button and D.A. Hensher (eds). Handbook of Transport Systems and Traffic Control. Pergamon, Amsterdam.

Madsen, H.O., Krenk, S. and Lind, N.C. (1985). Methods of Structural Safety. Prentice-Hall, Englewood Cliffs, NJ.

Madsen, H.O., Torhaug, R. and Cramer, E.H. (1991). Probability-based cost benefit analysis of fatigue design, inspection and maintenance. Proceedings of the Marine Structural Inspection, Maintenance and Monitoring Symposium, SSC/SNAME, Arlington, VA., II.E.1–12.

Maes, M. A., Fritzson, K. E. and Glowienka, S. (2006). Structural robustness in the light of risk and consequence analysis. *Structural Engineering International*, IABSE, 16(2): 101–107.

Mahmoud, H.N., Connor, R.J. and Bowman, C.A. (2005). Results of the Fatigue Evaluation and Field Monitoring of the I-39 Northbound Bridge over the Wisconsin River. ATLSS Rep. No. 05–04, Lehigh Univ., Bethlehem, PA.

Maljaars, J. and Vrouwenvelder, A.C.W.M. (2014). Probabilistic fatigue life updating accounting for inspections of multiple critical locations. *International Journal of Fatigue*, Elsevier, 68: 24–37.

Marelli, S. and Sudret, B. (2014). UQLab: A framework for uncertainty quantification in MATLAB. In Proceeding of the 2th International Conference on Vulnerability, Risk, Analysis and Management (ICVRAM2014), Liverpool, UK, 2014.

Marler, R.T. and Arora, J.S. (2010). The weighted sum method for multi-objective optimization: new insights. *Structural and Multidisciplinary Optimization*, 41(6): 853–862.

Marsh, P. S. and Frangopol, D. M. (2008). Reinforced concrete bridge deck reliability model incorporating temporal and spatial variations of probabilistic corrosion rate sensor data. *Reliability Engineering and System Safety*, Elsevier, 93(3): 364–409.

Martinez-Luengo, M., Kolios, A. and Wang, L. (2016). Structural health monitoring of offshore wind turbines: a review through the Statistical Pattern Recognition Paradigm. *Renewable and Sustainable Energy Reviews*, Elsevier, 64: 91–105.

Melchers, R.E. (1999). Structural Reliability Analysis and Prediction. 2nd Edition, John Wiley & Sons Ltd., U.K.

Messervey, T.B., Frangopol, D.M. and Casciati, S. (2011). Application of the statistics of extremes to the reliability assessment and performance prediction of monitored highway bridges. *Structure and Infrastructure Engineering*, Taylor & Francis, 7(1–2): 87–99.

Metropolis, N., Rosenbluth, A.W., Rosenbluth, M.N., Teller, A.H. and Teller, E. (1953). Equations of state calculations by fast computing machines. *Journal of Chemical Physics*, AIP, 21: 1087–1092.

Miner, M.A. (1945). Cumulative damage in fatigue. *Journal of Applied Mechanics*, ASME, 12(3): A159–A164.

Mirzaei, Z., Adey, B.T., Klatter, L. and Kong, J.S. (2012). The IABMAS Bridge Management Committee Overview of Existing Bridge Management Systems, Report by Bridge Management Committee of IABMAS.

Modarres, M., Kaminskiy, M.P. and Krivtsov, V. (2017) Reliability Engineering & Risk Analysis. 3rd Edition. CRC Press, U.S.

Mondoro, A., Frangopol, D.M. and Soliman, M. (2017). Optimal risk-based management of coastal bridges vulnerable to hurricanes. *Journal of Infrastructure Systems*, ASCE, 23(3): 4016046.

Mondoro, A. and Frangopol, D.M. (2018). Risk-based cost-benefit analysis for the retrofit of bridges exposed to extreme hydrologic events considering multiple failure modes. *Engineering Structures*, Elsevier, 159: 310–319.

Mondoro, A., Frangopol, D.M. and Liu, L. (2018a). Bridge adaptation and management under climate change uncertainties: a review. *Natural Hazards Review*, ASCE, 19(1): 4017023.

Mondoro, A., Frangopol, D.M. and Liu, L. (2018b). Multi-criteria robust optimization framework for bridge adaptation under climate change. *Structural Safety*, Elsevier, 74: 14–23.

Morgenthal, G. and Hallermann, N. (2014). Quality assessment of unmanned aerial vehicle (UAV) based visual inspection of structures. *Advances in Structural Engineering*, SAGE, 17(3): 289–302.

Mori, Y. and Ellingwood, B. (1993). Methodology for reliability-based condition assessment: Application to concrete structures in nuclear plants. NUREG/CR-6052, U.S. Nuclear Regulatory Commission, Washington, D.C.

Mori, Y. and Ellingwood, B.R. (1994a). Maintaining reliability of concrete structures. I: Role of inspection/repair. *Journal of Structural Engineering*, ASCE, 120(3): 824–845.

Mori, Y. and Ellingwood, B.R. (1994b). Maintaining reliability of concrete structures. II: Optimum inspection/repair. *Journal of Structural Engineering*, ASCE, 120(3): 846–862.

Myers, J.L., Well, A.D. and Lorch Jr., R.F. (2003). Research Design and Statistical Analysis. Routledge, NY.

Myung, H., Jeon, Y, Bang, Y.-S. and Wang, Y. (2014). Robotic sensing for assessing and monitoring civil infrastructures. Chapter 15. pp. 410–445. *In*: M.L. Wang, J.P. Lynch and H. Sohn (eds). Sensor Technologies for Civil Infrastructures, Vol. 1: Sensing Hardware and Data Collection Methods for Performance Assessment. Woodhead Publishing, Cambridge, U.K.

NCHRP (1997). Collecting and Managing Cost Data for Bridge Management Systems. Synthesis of Highway Practice 227, National Academy Press, Washington, D.C.

NCHRP (2003). Bridge life-cycle cost analysis. NCHRP-report 483, Transportation Research Board, Washington D.C.

NCHRP (2004). Traffic data collection, analysis, and forecasting for mechanic pavement design. NCHRP-report 538, Transportation Research Board, National Cooperative Highway Research Program, Washington D.C.

NCHRP (2005). Concrete bridge deck performance. NCHRP-synthesis 333, Transportation Research Board, National Cooperative Highway Research Program, Washington D.C.

NCHRP (2006). Manual on service life of corrosion-damaged reinforced concrete bridge superstructure elements. NCHRP-report 558, Transportation Research Board, National Cooperative Highway Research Program, Washington D.C.

NCHRP (2007). Countermeasures to Protect Bridge Abutments from Scour. NCHRP Report 587, Transportation Research Board, Washington, D.C.

NCHRP (2009). Best Practices in Bridge Management Decision-Making. Scan Team Report Scan 07–05, National Cooperative Highway Research Program (NCHRP), Washington, D.C.

Neal, R.M. (2003). Slice sampling. *Annals of Statistics*, IMS, 31(3): 705–767.

Neves, A.C. and Frangopol, D.M. (2005). Condition, safety and cost profiles for deteriorating structures with emphasis on bridges. *Reliability Engineering & System Safety*, Elsevier, 89(2): 185–198.

Neves, L.C., Frangopol, D.M. and Cruz, P.S. (2004). Cost of life extension of deteriorating structures under reliability-based maintenance. *Computers & Structures*, Elsevier, 82(13–14): 1077–1089.

Neves, L.A.C., Frangopol, D.M. and Cruz, P.J.S. (2006a). Probabilistic lifetime-oriented multi-objective optimization of bridge maintenance: single maintenance type. *Journal of Structural Engineering*, ASCE, 132(6): 991–1005.

Neves, L.A.C., Frangopol, D.M. and Petcherdchoo, A. (2006b). Probabilistic lifetime-oriented multi-objective optimization of bridge maintenance: combination of maintenance types. *Journal of Structural Engineering*, ASCE, 132(11): 1821–1834.

Ng, A.K.S. and Efstathiou, J. (2006). Structural robustness of complex networks. Proceedings of the international workshop and conference on network science (NetSci2006), May 16–25, 2006, Bloomington, Indiana.

Nickitopoulou, A., Protopsalti, K. and Stiros, S. (2006). Monitoring dynamic and quasi-static deformations of large flexible engineering structures with GPS: Accuracy, limitations and promises. *Engineering Structures*, Elsevier, 28(10): 1471–1482.

Nijkamp, P. and van Delft, A. (1977). Multi-Criteria Analysis and Regional Decision-Making. Springer Science & Business Media, Leiden, The Netherlands.

Nowak, A.S. (1999). Calibration of LRFD bridge design code. NCHRP-report 368. Transportation Research Council, Washington D.C.

NYSDOT (2008). Fundamentals of Bridge Maintenance and Inspection. Office of Transportation Maintenance, New York State Department of Transportation (NYSDOT), Albany, NY.

Okasha, N.M. and Frangopol, D.M. (2009a). Time-variant redundancy of structural systems. *Structure and Infrastructure Engineering*. Taylor & Francis, 6(1–2): 279–301.

Okasha, N.M. and Frangopol, D.M. (2009b). Lifetime-oriented multi-objective optimization of structural maintenance, considering system reliability, redundancy, and life-cycle cost using GA. *Structural Safety*, Elsevier, 31(6): 460–474.

Okasha, N.M. and Frangopol, D.M. (2010a). Redundancy of structural systems with and without maintenance: an approach based on lifetime functions. *Reliability Engineering & System Safety*, Elsevier, 95(5): 520–533.

Okasha, N.M. and Frangopol, D.M. (2010b). Novel approach for multicriteria optimization of life-cycle preventive and essential maintenance of deteriorating structures. *Journal of Structural Engineering*, ASCE, 136(8): 1009–1022.

Okasha, N.M. and Frangopol, D.M. (2010c). Advanced modeling for efficient computation of life-cycle performance prediction and service-life estimation of bridges. *Journal of Computing in Civil Engineering*, ASCE, 24(6): 548–556.

Okasha, N.M. and Frangopol, D.M. (2011). Computational platform for the integrated life-cycle management of highway bridges. *Engineering Structures*, Elsevier, 33(7): 2145–2153.

Okasha, N.M. and Frangopol, D.M. (2012). Integration of structural health monitoring in a system performance based life-cycle bridge management framework. *Structure and Infrastructure Engineering*, Taylor & Francis, 8(11): 999–1016.

Okasha, N.M., Frangopol, D.M. and Decò, A. (2010). Integration of structural health monitoring in life-cycle performance assessment of ship structures under uncertainty. *Marine Structures*, Elsevier, 23(3): 303–321.

Okasha, N.M., Frangopol, D.M. and Orcesi, A.D. (2012). Automated finite element updating using strain data for the lifetime reliability assessment of bridges. *Reliability Engineering & System Safety,* Elsevier, 99: 139–150.

Okasha, N.M., Frangopol, D.M., Saydam, D. and Salvino, L.W. (2011). Reliability analysis and damage detection in high speed naval crafts based on structural health monitoring data. *Structural Health Monitoring*, Sage Publication, 10(4): 361–379.

Okeil, A.M. and Cai, C.S. (2008). Survey of short-and medium- span bridge damage induced by Hurricane Katrina. *Journal of Bridge Engineering*, ASCE, 13(4): 377–387.

Orcesi, A.D. and Cremona, C.F. (2011). Optimal maintenance strategies for bridge networks using the supply and demand approach. *Structure and Infrastructure Engineering*, Taylor & Francis, 7(10): 765–781.

Orcesi, A.D. and Frangopol, D.M. (2010). Inclusion of crawl tests and long-term health monitoring in bridge serviceability analysis. *Journal of Bridge Engineering*, ASCE, 15(3): 312–326.

Orcesi, A.D. and Frangopol, D.M. (2011a). Optimization of bridge maintenance strategies based on structural health monitoring information. *Structural Safety*, Elsevier, 33(1): 26–41.

Orcesi, A.D. and Frangopol, D.M. (2011b). Probability-based multiple-criteria optimization of bridge maintenance using monitoring and expected error in the decision process. *Structural and Multidisciplinary Optimization*, Springer, 44(1): 137–148.

Orcesi, A.D. and Frangopol, D.M. (2011c). A stakeholder probability-based optimization approach for cost-effective bridge management under financial constraints. *Engineering Structures*, Elsevier, 33(5): 1439–1449.

Orcesi, A.D. and Frangopol, D.M. (2013). Bridge performance monitoring based on traffic data. *Journal of Engineering Mechanics*, ASCE, 139(11): 1508–1520.

Orcesi, A.D., Frangopol, D.M. and Kim, S. (2010). Optimization of bridge maintenance strategies based on multiple limit states and monitoring. *Engineering Structures*, Elsevier, 32(3): 627–640.

Ostachowicz, W., Soman, R. and Malinowski, P. (2019). Optimization of sensor placement for structural health monitoring: a review. *Structural Health Monitoring*, SAGE, 18(3): 963–988.

Packman, P.F., Pearson, H.S., Owens, J.S. and Young, G. (1969). Definition of fatigue cracks through nondestructive testing. *Journal of Materials*, ASTM, 4(3): 666–700.

Padgett, J.E., Dennemann, K. and Ghosh, J. (2010). Risk-based seismic life-cycle cost-benefit (LCC-B) analysis for bridge retrofit assessment. *Structural Safety*, Elsevier, 32(3): 165–173.

Padgett, J.E. and DesRoches, R. (2007). Bridge functionality relationships for improved seismic risk assessment of transportation networks. *Earthquake Spectra*, SAGE, 23(1): 115–130.

Pandey, A. K. and Biswas, M. (1995). Experimental verification of flexibility difference method for locating damage in structures. *Journal of Sound and Vibration*, Elsevier, 184(2): 311–328.

Paris P, and Erdogan F. (1963). A critical analysis of crack propagation laws. *Journal of Basic Engineering*, ASME, 85(4): 528–533.

Patelli, E. (2017). COSSAN: A multidisciplinary software suite for uncertainty quantification and risk management. pp. 1909–1977. *In*: R. Ghanem, D. Higdon and H. Owhadi (eds). Handbook of Uncertainty Quantification. Springer.

Peeters, B., Maeck, J. and Roeck, G.De (2001). Vibration-based damage detection in civil engineering: excitation sources and temperature effects. *Smart Materials and Structures*, IOP Publishing, 10(3): 518–527.

Perrings, C. (2006). Resilience and sustainable development. *Environment and Development Economics*, Cambridge University Press, 11(4): 417–427.

Quimpo, R.G. and Wu, S. (1997). Condition assessment of water supply infrastructure. *Journal of Infrastructure System*, ASCE, 3(1): 15–22.

Rackwitz, R. (1996). The program package STRUREL. pp. 360–367. *In*: Structural Reliability Methods. John Wiley, Chichester.

Rackwitz, R. (2006). The effect of discounting, different mortality reduction schemes and predictive cohort life tables on risk acceptability criteria. *Reliability Engineering & System Safety,* Elsevier, 91(4): 469–484

Rao, S.S. (2009). Engineering Optimization: Theory and Practice. 4th Edition. John & Wiley, N.J.

Rastogi, R., Ghosh, S., Ghosh, A.K., Vaze, K.K. and Singh, P.K. (2017). Fatigue crack growth prediction in nuclear piping using Markov chain Monte Carlo simulation. *Fatigue & Fracture of Engineering Materials & Structures*, Wiley Publishing Ltd., 40(1): 145–156.

Rausand, M. and Hoyland, A. (2004). System Reliability Theory: Models, Statistical Methods, and Applications, Wiley, New York.

Resnick, S.I. (1987). Extreme Values, Regular Variation and Point Processes. Springer-Verlag, New York.

Roberge, P.R. (1999). Handbook of Corrosion Engineering. McGraw-Hill, NewYork, USA.

Robert, C.P. and Cassella, G., (1999). Monte Carlo Statistical Methods. Springer Texts in Statistics. Springer-Verlag, New York.

Robert, C.P., (1994). The Bayesian Choice. Springer Texts in Statistics. Springer-Verlag, New York.

Rokneddin, K., Ghosh, J., Dueñas-Osorio, L. and Padgett, J.E. (2013). Bridge retrofit prioritization for ageing transportation networks subject to seismic hazards. *Structure and Infrastructure Engineering*, Taylor & Francis, 9(10): 1050–1066.

Rose, A. (2004). Defining and measuring economic resilience to disasters. *Disaster Prevention and Management*, Emerald Publishing, 13(4): 307–314.

Rosenzweig, C., Solecki, W.D., Blake, R., Bowman, M., Faris, C., Gornitz, V., Horton, R., Jacob, K., LeBlanc, A., Leichenko, R., Linkin, M., Major, D., O'Grady, M., Patrick, L., Sussman, E., Yohe, G. and Zimmerman, R. (2011). Developing coastal adaptation to climate change in the New York City infrastructure-shed: process, approach, tools, and strategies. *Climatic Change*, Springer, 106(1): 93–127.

Roy, B. (1971). Problems and methods with multiple objective functions. *Mathematical Programming*, Springer, 1(1): 239–266.

Sabatino, S. and Frangopol, D.M. (2017). Decision making framework for optimal SHM planning of ship structures considering availability and utility. *Ocean Engineering*, Elsevier, 135: 194–206.

Safi, M., Sundquist, H., Karoumi, R. and Racutanu, G. (2013). Development of the Swedish bridge management system by upgrading and expanding the use of LCC. *Structure and Infrastructure Engineering*. Taylor & Francis, 9(12): 1240–1250.

Sagar, R.V. and Dutta, M. (2021). Combined usage of acoustic emission technique and ultrasonic pulse velocity test to study crack classification in reinforced concrete structures. *Nondestructive Testing and Evaluation*, Taylor & Francis, 36(1): 62–96.

Sánchez-Silva, M., Frangopol, D.M., Padgett, J. and Soliman, M. (2016). Maintenance and operation of infrastructure systems: review. *Journal of Structural Engineering*, ASCE, 142(9): F4016004.

Sawyer, A.D. (2008). Determination of Hurricane Surge Wave Forces on Bridge Superstructures and Design/Retrofit Options to Mitigate or Sustain These Forces. M.S. thesis, Auburn Univ., Auburn, AL.

Saxena, D.K., Duro, J.A., Tiwari, A., Deb, K. and Qingfu Zhang (2013). Objective reduction in many-objective optimization: linear and nonlinear algorithms. *Evolutionary Computation*, IEEE, 17(1): 77–99.

Saydam, D. and Frangopol, D. M. (2011). Time-dependent performance indicators of damaged bridge superstructures. *Engineering Structures*, Elsevier, 33(9): 2458–2471.

Saydam, D., Bocchini, P. and Frangopol, D.M. (2013a). Time-dependent risk associated with deterioration of highway bridge networks. *Engineering Structures*, Elsevier, 54: 221–233.

Saydam, D., Frangopol, D.M. and Dong, Y. (2013b). Assessment of risk using bridge element condition ratings. *Journal of Infrastructure Systems*, ASCE, 19(3): 252–265.

Schijve, J. (2003). Fatigue of structures and materials in the 20th century and the state of the art. *International Journal of Fatigue*, Elsevier, 25(8): 679–702.

Scott, D.M., Novak, D.C., Aultman-Hall, L. and Guo, F. (2006). Network robustness index: a new method for identifying critical links and evaluating the performance of transportation networks. *Journal of Transport Geography*, Elsevier, 14(3): 215–227.

Scutaru, M.C., Comisu, C.C., Boacă, G. and Țăranu, N. (2018). Bridge maintenance strategies – A brief comparison among different countries around the world. Proceedings of the Ninth International Conference on Bridge Maintenance, Safety and Management, July 9–13, 2018, Melbourne, Australilia.

Seo, J., Hu, J.W. and Lee, J. (2016). Summary review of structural health monitoring applications for highway bridges. *Journal of Performance of Constructed Facilities*, ASCE, 30(4): 4015072.

Shafei, B., Alipour, A. and Shinozuka, M. (2012). Prediction of corrosion initiation in reinforced concrete members subjected to environmental stressors: a finite-element framework. *Cement Concrete Research*, Elsevier, 42(2): 365–376.

Shaw, R.S. (2002). Ultrasonic testing procedures, technician skills, and qualifications. *Journal of Materials in Civil Engineering*, ASCE, 14(1): 62–67.

Shields, M.D., Giovanis, D.G., Satish, A.B., Chauhan, M.S., Olivier, A., Vandanapu, L., Zhang, J. (2019). Uncertainty quantification with python (UQpy). https://github.com/SURGroup/UQpy.

Singh, H.K., Isaacs, A. and Ray, T. (2011). A Pareto corner search evolutionary algorithm and dimensionality reduction in many-objective optimization problems. *Evolutionary Computation*, IEEE Transactions on, 15(4): 539–556.

Soliman, M. and Frangopol, D.M. (2014). Life-cycle management of fatigue sensitive structures integrating inspection information. *Journal of Infrastructure Systems*, ASCE, 20(2): 04014001, 1–13.

Soliman, M. and Frangopol, D.M. (2015). Life-cycle cost evaluation of conventional and corrosion-resistant steel for bridges. *Journal of Bridge Engineering*, ASCE, 20(1): 0614005.

Soliman, M., Frangopol, D.M. and Kim, S. (2013a). Probabilistic optimum inspection planning of steel bridges with multiple fatigue sensitive details. *Engineering Structures*, Elsevier, 49: 996–1006.

Soliman, M., Frangopol, D.M. and Kwon, K. (2013b). Fatigue assessment and service life prediction of existing steel bridges by integrating SHM into a probabilistic bi-linear S-N approach. *Journal of Structural Engineering*, ASCE, 139(10): 1728–1740.

Soliman, M., Frangopol, D.M. and Mondoro, A. (2016). A probabilistic approach for optimizing inspection, monitoring, and maintenance actions against fatigue of critical ship details. *Structural Safety*, Elsevier, 60: 91–101.

Sprinkel, M.M. (2001). Maintenance of concrete bridges. *Transportation Research Record: Journal of Transportation Research Board*, SAGE, 1749(1): 60–63.

SRA (1996a). Bridge—Inspection Manual. Swedish Road Administration (SRA) publication 1996:036(E), Borlänge, Sweden.

SRA (1996b). Bridge—Measurement and Condition Assessment. Swedish Road Administration (SRA) publication 1996:038(E), Borlänge, Sweden.

SRA (2002). SAFEBRO—Koder för inspektion. Swedish Road Administration (SRA) publication 2002:77, Borlänge, Sweden.

Stein, S., Young, G.K., Trent, R.E. and Pearson, D.R. (1999). Prioritizing scour vulnerable bridges using risk. *Journal of Infrastructure Systems*, ASCE, 5(3): 95–101.

Stewart, M.G. (2004). Spatial variability of pitting corrosion and its influence on structural fragility and reliability of RC beams in flexure. *Structural Safety*, Elsevier, 26(4): 453–470.

Stewart, M.G. and Deng, X. (2015). Climate impact risks and climate adaptation engineering for built infrastructure. *ASCE-ASME Journal of Risk and Uncertainty in Engineering Systems, Part A: Civil Engineering*, ASCE, 1(1): 4014001.

Stewart, M.G. and Mueller, J. (2014). Terrorism risks for bridges in a multi-hazard environment. *International Journal of Protective Structures*, SAGE, 5(3): 275–289.

Stewart, M.G. and Rosowsky, D.V. (1998a). Time-dependent reliability of deteriorating reinforced concrete bridge decks. *Structural Safety*, Elsevier, 20(1): 91–109.

Stewart, M.G. and Rosowsky, D.V. (1998b). Structural safety and serviceability of concrete bridges subject to corrosion. *Journal of Infrastructure Systems*, ASCE, 4(4):146–155.

Stewart, M.G., Estes, A.C. and Frangopol, D.M. (2004). Bridge deck replacement for minimum expected cost under multiple reliability constraints. *Journal of Structural Engineering*, ASCE, 130(9): 1414–1419.

Stewart, M.G., Wang, X. and Nguyen, M.N. (2011). Climate change impact and risks of concrete infrastructure deterioration. *Engineering Structures*, Elsevier, 33(4): 1326–1337.

Stewart, M.G., Wang, X. and Nguyen, M.N. (2012). Climate change adaptation for corrosion control of concrete infrastructure. *Structural Safety*, Elsevier, 35: 29–39.

Straub, D. (2009). Stochastic modeling of deterioration processes through dynamic Bayesian networks. *Journal of Engineering Mechanics*, ASCE, 135(10): 1089–1099.

Strauss, A., Frangopol, D.M. and Kim, S. (2008). Use of monitoring extreme data for the performance prediction of structures: Bayesian updating. *Engineering Structures*, Elsevier, 30(12): 3654–3666.

Streicher, H., Joanni, A. and Rackwitz, R. (2008). Cost-benefit optimization and risk acceptability for existing, aging but maintained structures. *Structural Safety*, Elsevier, 30(5): 375–393.

Suo, Q. and Stewart, M.G. (2009). Corrosion cracking prediction updating of deteriorating RC structures using inspection information. *Reliability Engineering & System Safety*, Elsevier, 94(8): 1340–1348.

Ta, M.-N., Lardis, J. and Marc, B. (2006). Natural frequencies and modal damping ratios identification of civil structures from ambient vibration data. *Shock and Vibration*, IOS Press, 13(4–5): 429–444.

Thanapol, Y., Akiyama, M. and Frangopol, D.M. (2016). Updating the seismic reliability of existing RC structures in a marine environment by incorporating the spatial steel corrosion distribution: Application to bridge piers. *Journal of Bridge Engineering*, ASCE, 21(7): 04016031, 1–17.

Thoft-Christensen, P. (2003). Corrosion and Cracking of Reinforced Concrete. pp. 26–36. *In*: D.M. Frangopol, E. Brühwiler, M.H. Faber and B. Adey (eds). Life-Cycle Performance of Deteriorating Structures. ASCE, Reston, Virginia.

Thoft-Christensen, P. (2009). Life-cycle cost-benefit (lccb) analysis of bridges from a user and social point of view. *Structure and Infrastructure Engineering*, Taylor & Francis, 5(1): 49–57.

Tierney, K. and Bruneau, M. (2007). Conceptualizing and Measuring Resilience. *TR News*, No. 250: 14–17, Transportation Research Board, Washington, D.C.

Tierney, L. and Mira, A. (1999). Some adaptive Monte Carlo methods for Bayesian inference. *Statistics in Medicine*, Wiley & Sons, 18(17–18): 2507–2515.

TRB (2000). Highway capacity manual. Special Report 209, Transportation Research Board (TRB), National Research Council, Washington, D.C.

Val, D.V. and Melchers, R.E. (1997). Reliability of deteriorating RC slab bridges. *Journal of Structural Engineering*, ASCE, 123(12): 1638–1644.

Val, D.V. and Stewart, M.G. (2003). Life-cycle cost analysis of reinforced concrete structures in marine environments. *Structural Safety*, Elsevier, 25(4): 343–362.

van Noortwijk, J.M. and Frangopol, D.M. (2004). Two probabilistic life-cycle maintenance models for deteriorating civil infrastructures. *Probabilistic Engineering Mechanics*, Elsevier, 19(4): 345–359.

Velasquez, M. and Hester, P.T. (2013). An analysis of multi-criteria decision making methods. *International Journal of Operations Research*, ORSTW, 10(2): 56–66.

Verma, S.K., Bhadauria, S.S. and Akhtar, S. (2013). Review of nondestructive testing methods for condition monitoring of concrete structures. *Journal of Construction Engineering*, Hindawi, 2013: 834572.

Voogd, H. (1983). Multicriteria Evaluation for Urban and Regional Planning. Pion, London.

Vu, K.A.T. and Stewart, M.G. (2000). Structural reliability of concrete bridges including improved chloride-induced corrosion models. *Structural Safety*, Elsevier, 22(4): 313–333.

Wang, X., Stewart, M.G. and Nguyen, M. (2012). Impact of climate change on corrosion and damage to concrete infrastructure in Australia. *Climatic Change*, Springer, 110(3): 941–957.

Washer, G. (2014). Nondestructive evaluation methods for bridge elements. Chapter 12. pp. 301–336. *In*: W-F. Chen and L. Duan (eds). Bridge Engineering Handbook, 2nd Edition, Vol. 5: Construction and Maintenance. CRC Press/Taylor & Francis Group, Boca Raton, London, New York.

Washer, G., Connor, R. and Looten, D. (2014). Performance testing of inspectors to improve the quality of nondestructive testing. *Transportation Research Record*, SAGE, 2408(1): 107–113.

Xu, Y. and Brownjohn, J.M.W. (2018). Review of machine-vision based methodologies for displacement measurement in civil structures. *Journal of Civil Structural Health Monitoring*, Springer, 8(1): 91–110.

Yadollahi, M., Ansari, R., Abd Majid, M.Z. and Yih, C.H. (2015). A multi-criteria analysis for bridge sustainability assessment: a case study of Penang Second Bridge, Malaysia. *Structure and Infrastructure Engineering*, Taylor & Francis, 11(5): 638–654.

Yan, L. and Frangopol, D.M., (2019). Utility and information analysis for optimum inspection of fatigue-sensitive structures. *Journal of Structural Engineering*, ASCE, 145(2): 04018251.

Yanev, B. (2003). Management for the bridges of New York City. *International Journal of Steel Structures*, Korean Society of Steel Construction (KSSC), 3(2): 127–35.

Yang, D.Y. and Frangopol, D.M. (2018a). Bridging the gap between sustainability and resilience of civil infrastructure using lifetime resilience. Chapter 23. pp. 419–442. *In*: P. Gardoni (ed.) Routledge Handbook of Sustainable and Resilient Infrastructure. Routledge.

Yang, D.Y. and Frangopol, D.M. (2018b). Risk-informed bridge ranking at project and network levels. *Journal of Infrastructure Systems*, ASCE, 24(3): 4018018.

Yang, D.Y. and Frangopol, D.M. (2018c). Probabilistic optimization framework for inspection/repair planning of fatigue-critical details using dynamic Bayesian networks. *Computers and Structures*, Elsevier, 198: 40–50.

Yang, D.Y. and Frangopol, D.M. (2019a). Life-cycle management of deteriorating civil infrastructure considering resilience to lifetime hazards: A general approach based on renewal-reward processes. *Reliability Engineering and System Safety*, Elsevier, 183: 197–212.

Yang, D.Y. and Frangopol, D.M. (2019b). Physics-based assessment of climate change impact on long-term regional bridge scour risk using hydrologic modeling: application to Lehigh river watershed. *Journal of Bridge Engineering*, ASCE, 24(11): 4019099.

Yang, D.Y. and Frangopol, D.M. (2019c). Societal risk assessment of transportation networks under uncertainties due to climate change and population growth. *Structural Safety*, Elsevier, 78: 33–47.

Yang, D.Y. and Frangopol, D.M. (2020a). Life-cycle management of deteriorating bridge networks with network-level risk bounds and system reliability analysis. *Structural Safety*, Elsevier, 83: 101911.

Yang, D.Y. and Frangopol, D.M. (2020b). Risk-based vulnerability analysis of deteriorating coastal bridges under hurricanes considering deep uncertainty of climatic and socioeconomic changes. *ASCE-ASME Journal of Risk and Uncertainty in Engineering Systems, Part A: Civil Engineering*, ASCE, 6(3): 04020032.

Yang, D.Y. and Frangopol, D.M. (2021). Risk-based inspection planning of deteriorating structures. *Structure and Infrastructure Engineering*, Taylor & Francis, https://doi.org/10.1080/15732479.2021.1907600 (published online: May 3, 2021), 1–20.

Yang, S.-I., Frangopol, D.M. and Neves, L.C. (2004). Service life prediction of structural systems using lifetime functions with emphasis on bridges. *Reliability Engineering and System Safety*, Elsevier, 86(1): 39–51.

Yang, S.-I., Frangopol, D.M. and Neves, L.C. (2006a). Optimum maintenance strategy for deteriorating bridge structures based on lifetime functions. *Engineering Structures*, Elsevier, 28(2): 196–206.

Yang, S.-I., Frangopol, D.M., Kawakami, Y. and Neves, L.C. (2006b). The use of lifetime functions in the optimization of interventions on existing bridges considering maintenance and failure costs. *Reliability Engineering & System Safety*, Elsevier, 96(6): 698–705.

Yari, N. (2018). New Model for Bridge Management System (BMS): Bridge Repair Priority Ranking System (BRPRS), Case Based Reasoning for Bridge Deterioration, Cost Optimization, and Preservation Strategy, Ph.D. dissertation, University of New Hampshire, Durham, NH.

Yeh, C.-H. (2002). A problem-based selection of multi-attribute decision-making methods. *International Transactions in Operational Research*, Blackwell Publishing, 9(2): 169–181.

Yeo, G.L. and Cornell, C.A. (2005). Stochastic characterization and decision bases under time-dependent aftershock risk in performance-based earthquake engineering. PEER Rep. 2005/13, Univ. of California, Berkeley, CA.

Yi, T., Li, H. and Gu, M. (2010). Recent research and applications of GPS based technology for bridge health monitoring. *Science China Technological Sciences*, Springer, 53(10): 2597–2610.

Yoon, H., Shin, J. and Spencer Jr, B.F. (2018). Structural displacement measurement using an unmanned aerial system. *Computer-Aided Civil Infrastructure Engineering*, John Wiley & Sons Ltd, 33: 183–192.

Yoon, K.P. and Hwang, C.-L. (1981). Multiple Attribute Decision Making: Methods and Applications. Springer: Berlin/Heidelberg, Germany.

Yoon, K.P. and Hwang. C.-L. (1995). Multiple Attribute Decision Making: An Introduction. SAGE Publication Inc., London.

Zanakis, S.H., Solomon, A., Wishart, N. and Dublish, S. (1998). Multi-attribute decision making: a simulation comparison of select methods. *European Journal of Operational Research*, Elsevier, 107(3): 507–529.

Zhang, J. and Lounis, Z. (2006). Sensitivity analysis of simplified diffusion-based corrosion initiation model of concrete structures exposed to chlorides. *Cement and Concrete Research*, Elsevier, 36(7): 312–323.

Zhang, W. and Wang, N. (2017). Bridge network maintenance prioritization under budget constraint. *Structural Safety*, Elsevier, 67: 96–104.

Zhang, W., Cai, C.S., Pan, F. and Zhang, Y. (2014). Fatigue life estima- tion of existing bridges under vehicle and non-stationary hurricane wind. *Journal of Wind Engineering & Industrial Aerodynamics*, Elsevier, 133: 135–145.

Zhu, B. and Frangopol, D.M. (2012). Reliability, redundancy and risk as performance indicators of structural systems during their life-cycle. *Engineering Structures*, Elsevier, 41(8): 34–49.

Zhu, B. and Frangopol, D.M. (2013a). Risk-based approach for optimum maintenance of bridges under traffic and earthquake loads. *Journal of Structural Engineering*, ASCE, 139(3): 422–434.

Zhu, B. and Frangopol, D.M. (2013b). Reliability assessment of ship structures using Bayesian updating. *Engineering Structures*, Elsevier, 56: 1836–1847.

Zhu, B. and Frangopol, D.M. (2013c). Incorporation of SHM data on load effects in the reliability and redundancy assessment of ships using Bayesian updating. *Structural Health Monitoring*, SAGE, 12(4): 377–392.

Zhu, B. and Frangopol, D.M. (2016). Time-dependent risk assessment of bridges based on cumulative-time failure probability. *Journal of Bridge Engineering*, ASCE, 21(12): 06016009, 1–7.

Ziehl, P., Galati, N., Tumialan, G. and Nanni, A. (2008). In-situ evaluation of two concrete slab systems. II: evaluation criteria and outcomes. *Journal of Performance of Constructed Facilities*, ASCE, 22(4): 217–227.

Zoubir, L. and McAllister, T.P. (2016). Risk-based decision making for sustainable and resilient infrastructure systems. *Journal of Structural Engineering*, ASCE, 142(9): F4016005, 1–14.

Index